THE LAST DAYS OF STEAM
— AROUND —
NOTTINGHAM
AND DERBY

Arnold Road 64827
Gresley's Class J39 0-6-0 locomotive no. 64827 trundles through Arnold Road, Nottingham, on 20 September 1958. Built at Darlington the locomotive entered service in December 1931 and was withdrawn from Doncaster shed in February 1960. Bill was often able to get lineside for a number of his pictures in the 1950s and 1960s, never encountering any problems from the railway authorities.

New Basford 45280
For this picture, taken on 9 May 1964 at New Basford, Bill is standing on Perry Road bridge looking south towards Nottingham. Depicted is 'Black' Five 4-6-0 locomotive no. 45280, built at Armstrong Whitworth, and entering service in November 1936. At the time of the picture the engine was allocated to Saltley. Withdrawal was from Birkenhead Mollington Street in November 1967.

THE LAST DAYS OF STEAM
— AROUND —
NOTTINGHAM AND DERBY
FROM THE BILL REED COLLECTION

PETER TUFFREY

F

FONTHILL

Pictured in Derby Works Paint Shop on 14 May 1956 is Fowler 4F 0-6-0 locomotive no. 44518. The engine was erected at Crewe Works during July 1928 and at the time of the photograph was allocated to Peterborough Spital Bridge. No. 44518 was in service over thirty-four years before being condemned in September 1962.

I have taken reasonable steps to verify the accuracy of the information in this book but it may contain errors or omissions. Any information that may be of assistance to rectify any problems will be gratefully received. Please contact me by email petertuffrey@rocketmail.com or in writing: Peter Tuffrey, 8 Wrightson Avenue, Warmsworth, Doncaster, South Yorkshire, DN4 9QL.

Fonthill Media Limited
www.fonthillmedia.com
office@fonthillmedia.com

First published in the United Kingdom 2014

British Library Cataloguing in Publication Data
A catalogue record for this book is available from the British Library

ISBN 978-1-78155-056-4

Typeset in 10.5pt on 12pt Sabon LT Std
Printed and bound in England

Contents

This book is dedicated to the memory of George William Chambers who was better known simply as 'Pip.' He was born on 23 October 1907 and in time served on the NUR Committee, spending four years at Unity House in London. Returning to the footplate he subsequently became President of the Union. On 2 September 1967 he took the last train from Nottingham Victoria (to Rugby) before the station was closed. In later years George served on Nottingham City Council and County Council, becoming chairman of the latter in 1984. He died in his ninety-fifth year on 24 May 2002.

Bill Reed pictured alongside Class 37 locomotive no. 37 140.

Introduction

Alan Sutton, Fonthill Media MD, asked if I could delve once more into Bill Reed's massive photo collection to produce another book – *Last Days of Steam Around Nottingham and Derby* – and I was happy to oblige. The book is also a fitting tribute to Nottingham-born Bill as, at the time of writing, he has just celebrated his eightieth birthday.

The earliest pictures in the book date from 1949 when Bill, armed with a second-hand Kodak 127 camera, toured the Nottingham and Derby areas. Initially he collected locomotive numbers but once the photographic bug took hold he was more content taking pictures of the engines themselves.

Examples of his early photograph efforts may be seen at Colwick and Derby. It is also remarkable to hear that at the age of sixteen, Bill was never challenged about being on railway property. 'Nobody bothered in those days,' recalls Bill. 'I always behaved responsibly, wandered wherever I pleased, took pictures and cleared off. No trouble at all.'

At this time Bill says it was unthinkable that some of his pictures would one day find their way into a book as there were certainly not the same outlets or interest then as there is today. He also recalls that obtaining photographic materials was relatively easy though there were only a few railway photographers about at this period.

Bill acquired some of his photograph skills from well-known local photographer Freddie Guildford who gave him instructions on how to develop films and print photographs. Bill mostly undertook this work in a makeshift darkroom at his parent's house in Bulwell. It is worth noting that Freddie Guildford's own pictures sometimes appear in a number of nostalgic steam publications.

Initially, Bill found employment as a messenger lad at Nottingham Victoria station enquiry office and spent his lunchtime hours observing locomotive movements in and out of the station. By 1950, Bill had left his job at Nottingham Vic and went to the Midland Region at Middle Furlong Road to start work as an engine cleaner. This he hoped would set him on the long road to become a steam locomotive driver.

An intervening period in the early 1950s saw Bill undertaking National Service though once back in the UK he wasted no time obtaining an Agfa Isolette and shortly afterwards an Agfa Super Isolette camera on which he took many of the black and white photographs in the book which stretch over the period 1954 to 1966. He took pictures, sometimes accompanied by another local photographer,

Don Beecroft, in the immediate locality and often further afield travelling on the back of his friend's motorcycle.

Some of Bill's favourite local locations included vantage points at New Basford, Arnold Road, Bulwell Common, Bulwell Hall, the 'Rat Hole,' Toton, Watnall and Derby.

'On Sunday mornings', recalled Bill, 'I regularly visited Annesley shed then Kirkby and either Toton or Colwick in the afternoon.'

The pictures taken at Bulwell Common are from the bottom of his dad's allotment at the junction of Hucknall Road and St Alban's Road; the site now redeveloped beyond recognition. This can be said of many of the locations where Bill took his pictures none more so than the view from Perry Road Bridge looking down on to New Basford. Today the even the bridge is non-existent and high fences obliterate the once perfect view down on to the railway tracks. Houses also stretch across the in-filled cutting.

In time Bill became a fireman on steam locomotives but sadly only became a driver after steam's demise. Nevertheless, he fired many times for legendary driver George Chambers who served on the NUR National Committee and later became Nottinghamshire County Council chairman. On Bill's request this book is dedicated to George's memory. 'George was a good friend,' admitted Bill, 'and not like other drivers. He didn't frown upon me being a railway enthusiast. Other drivers would lead you a dog's life if they knew you had an interest in the job.'

Later on Bill was often called upon to take pictures of various railway functions and events as well as football teams some of which are included here.

Bill took a few pictures of a curious Nottingham location – Basford Tunnel – locally known as the 'Rat Hole.' This was part of a spur, from the former GN line, which dived underneath the latter and re-emerged to join the GC line at Bagthorpe Junction. 'In those days,' said Bill 'exposures etc were largely guess work and pictures of locos entering the 'Rat Hole' really tested my skills.'

In writing the captions I have tried to give brief histories of locations – shed and stations – as well as detailed information about a particular locomotive's build dates, allocations, alterations and withdrawal. For this a great number of reference sources have been used as will be deduced from the Bibliography at the end of the book.

Acknowledgements

I am grateful for the assistance of the following people: Iris Chambers, Hugh Parkin, Bill Reed and Alan Sutton.

Special thanks are due to my son Tristram Tuffrey for his general help and advice.

CHAPTER ONE

Annesley

Annesley 46156
Annesley shed was opened during 1898 by the Great Central Railway to the south of Annesley North Junction and Annesley exchange sidings, which was a hub for the area's coal traffic. The shed comprised six tracks that were enclosed by brick walls and covered by a slated roof. On the scrap line at Annesley, with at least two other London Midland & Scottish Railway Fowler Class 6 'Royal Scot' 4-6-0s, is no. 46156 *The South Wales Borderer*. During the British Railways period the class members were latecomers to the shed, with the first arriving in September 1962. This locomotive had resided at the shed since October 1963 and was withdrawn in October of the following year.

Annesley 46165

This Royal Scot's fate has been marked on its tender, bringing to an end a career that had lasted just over thirty-four years. The picture dates from 3 April 1965 and although withdrawal had occurred in November of the previous year, no. 46165 *The Ranger (12th London Regt.)* awaits its final journey to T. W. Ward Ltd, Beighton for disposal. Classmate no. 46125 *3rd Carabinier* can be seen in the background at the front of no. 46165. Annesley shed closed on 3 January 1966 and had been demolished by 1967.

Arnold Road

Arnold Road 44693

Just north of Bagthorpe Junction and heading into Nottingham on the former Great Central main line is Class Five 4-6-0 no. 44693, which has been pictured on 20 September 1958. Bagthorpe Junction was the southern link between the former Great Northern Railway line at Basford East Junction and the GCR main line. The locomotive was constructed at Horwich Works in November 1950 and was withdrawn from Low Moor shed, Bradford, in May 1967.

Arnold Road 44951

Travelling northwards on 2 March 1963 is LMS Stanier Class Five 4-6-0 no. 44951. The locomotive entered traffic from Horwich Works during March 1946 with boiler no. 12242, which had a 28-element superheater and sloping firebox throatplate. The boiler top feed has subsequently been altered from the original type to the new style, which had setscrews protruding from the centre and required a 'top hat' covering. No. 44951 left service in December 1966.

Arnold Road 45342

This Armstrong Whitworth-built Class Five or 'Black Five', no. 45342, has been photographed on 5 August 1964 when it was a resident of Annesley shed; the allocation lasted from July 1964 to June 1965. An earlier type of top feed with side-mounted setscrews is in evidence here, as is a buffer beam with flush rivets, top lamp iron moved to a lower position on the smokebox door and washout plugs on the first boiler ring. The locomotive was withdrawn in August 1968.

Arnold Road 61085

Running tender first towards Bagthorpe Junction during 1958 is LNER Thompson B1 Class 4-6-0 locomotive, no. 61085, which was erected by the North British Locomotive Company's Queens Park Works in October 1946. The engine is paired with a welded Group Standard 4,200 gallon tender that also had a coal capacity of 7 tons 10 cwt. Another feature visible here are the guard irons at the rear of the tender, which had been omitted from tenders constructed for the initial B1 Class locomotives. No. 61085 was condemned in November 1961 and later scrapped by Darlington Works.

Arnold Road 61094

No. 61094 was constructed by the NBLC a month after no. 61085 and both were built as part of an order placed at the company by the LNER during August 1945 for 100 engines. They were given LNER numbers 1040-1139 and all were completed between April 1946 and April 1947. When this image was captured on 2 March 1963 the locomotive was allocated to Sheffield shed, but by June no. 61094 would be transferred to Canklow. Colwick would later be visited briefly before withdrawal occurred during June 1965.

Arnold Road 61152

Another Sheffield-allocated B1 Class locomotive has been photographed by Bill near Bagthorpe Junction, but this picture dates from 20 September 1958. No. 61152 was built by the Vulcan Foundry, Newton-le-Willows, in May 1947 and then sent to Gorton, which along with Sheffield Darnall shed, was the focal point for deliveries of the Vulcan Foundry batch of B1 locomotives. The engine arrived at Sheffield in June 1947 and spent the majority of the next sixteen years working from the area, this only being broken by a two-month spell at Doncaster in late 1957.

Arnold Road 61209

B1 no. 61209 has just passed over the former GNR Derbyshire & Staffordshire Extension line. The Act for the route, between Colwick Junction, Derby and Burton, was passed on 25 July 1872 and the first stage to Pinxton was open for mineral traffic from 23 August 1875. The first passenger service between Nottingham and New Basford ran on 1 February 1876, but it was not until 1 April 1878 that passenger services from Nottingham to Derby were operational on the line. No. 61209 was erected by the NBLC in July 1947 and left service in September 1962.

Arnold Road 45562

LMS Stanier Class 5XP 'Jubilee' 4-6-0 locomotive no. 45562 *Alberta* was constructed at the NBLC's Hyde Park Works in July 1934 with a nickel steel boiler that had 160 small tubes and 14 superheater elements. Another a feature of this batch built by the NBLC was a 2 inch longer bogie wheelbase from standard at 6ft. 6in. and larger diameter bogie wheels that measured 3ft. 3½in. diameter. No. 45562's boiler was subsequently switched to one which had an improved arrangement with 159 small tubes and 24 superheater elements in July 1937.

Arnold Road 60908

LNER Gresley V2 Class 2-6-2 no. 60908 (from November 1948) emerged from Darlington Works in April 1940 as LNER no. 4879. A year before this picture was taken on 28 August 1954, the locomotive was employed on the Southern Region from Nine Elms shed due to the temporary withdrawal of the 'Merchant Navy' and 'Light' Pacific locomotives. This sojourn lasted from 14 May to 28 June 1953 and the engine then returned to Peterborough New England shed, which was its home from 1946 to 1959. Withdrawal occurred in June 1962 and no. 60908 was scrapped at Doncaster.

Arnold Road 60901

Thirty-six V2 Class locomotives were ordered from Darlington Works by the LNER during November 1938. They entered traffic between October 1939 and August 1940 taking numbers 4853-4888; no. 4872 (60901 from December 1949) was built in March 1940. The engine was a long time resident in the northeast and at the time of this picture, 28 August 1954, was allocated to York. June 1965 saw no. 60901 leave service.

Arnold Road 46128

Running light engine on 2 March 1963 is Royal Scot no. 46128 *The Lovat Scouts*. The locomotive was erected by the NBLC's Hyde Park Works in August 1927 at a cost to the LMS of £7,744. By April of the following year the engine had acquired the name *Meteor* and this was carried until November 1936. No. 46128 was allocated to Carlisle Upperby shed at the time of the picture and left service from there in May 1965.

Arnold Road 61248

Built in October 1947 by the NBLC's Queens Park Works, B1 Class locomotive no. 61248 was named *Geoffrey Gibbs* in December after one of the Directors on the LNER Board. Seventeen other B1 engines were similarly named in late 1947 before BR absorbed the company. No. 61248 has been fitted with Automatic Warning System apparatus, which had been added to the locomotive in June 1959 during a General repair at Stratford Works. This image dates from 2 March 1963 when *Geoffrey Gibbs* was allocated to Immingham.

Arnold Road 61821

Looking northwards along the former GCR main line, with Orville Road, Old Basford, in the background, LNER Gresley K3 Class 2-6-0 locomotive no. 61821 is seen approaching with a local passenger service on 20 September 1958. The engine was constructed at Darlington Works in October 1924 as part of the 17 series, which formed the basis of K3 Class part two; these locomotives being slightly altered from the first ten built at Doncaster before Grouping. No. 61821 was a long-term Colwick resident, but moved to Immingham two months before it was withdrawn in September 1962.

Arnold Road 61836

Another K3/2 Class locomotive is seen from the same position, but this picture dates from 11 June 1957. No. 61836 was completed in January 1925 as the second engine of the second batch of the 17 series, which had been ordered during November 1923 – a month after the first batch. The locomotives of the 17 series were built to the LNER Loading Gauge, had North Eastern Railway-style chimneys and Group Standard tenders. No. 61836 was condemned at Doncaster shed in February 1960.

Arnold Road 63594

No. 63954 was manufactured at Gorton Works in November 1911 and originally belonged to the GCR's 8K Class (later LNER O4/1 Class), which were designed by John G. Robinson for heavy goods and mineral traffic. Edward Thompson produced plans for a standard 2-8-0 to replace a number of existing classes and rebuilt a number of the O4 Class to the new design and they then became the O1 Class. No. 63954 was converted in November 1947 and ran in this form until it left service in April 1964. This photograph dates from 25 September 1957.

Arnold Road 62670

No. 62670 *Marne* is another former GCR locomotive, although it was only briefly a member of the Robinson 11F Class before they were reclassified D11/1 by the LNER after Grouping. The engine was also built at Gorton, leaving the works in December 1922 to begin service from Neasden. When the locomotive was pictured at Bagthorpe Junction on 20 September 1958 it was working from Sheffield Darnall shed, which used the class on a variety of stopping services. Withdrawal occurred in November 1960.

Arnold Road 64406

No. 64406 started life as part of the numerous GCR 9J Class of 0-6-0 goods locomotives designed by Robinson. The engine was produced at Gorton Works in April 1907 and originally it had a saturated boiler and slide valves. The locomotive later acquired an 18-element Robinson superheater in April 1935 as part of a class-wide modification started by the GCR in 1913. Thompson rebuilt a number of 9J locomotives (LNER Class J11) with piston valves in the 1940s and no. 64406 was altered in October 1943.

Arnold Road 63780

No. 63780 was one of 48 locomotives purchased by the LNER from the government in 1925 (entering traffic in July) and it was subsequently classified O4/3. The engine had been built to the design of Robinson's GCR 8K Class by the NBLC's Hyde Park Works in May 1918 for the Railway Operating Division of the Royal Engineers to use during the First World War. Thompson also rebuilt this locomotive to O1 Class specifications and the transformation occurred in April 1945. No. 63780 had recently relocated to March shed before Bill photographed the engine on 14 July 1957 and it was withdrawn from there in July 1963.

Arnold Road 73170

The penultimate BR Standard Class Five 4-6-0, no. 73170, passes by with an express passenger working on 2 March 1963. The locomotive was built at Doncaster in May 1957 and had a disappointingly short service life of just under ten years, being withdrawn in June 1966. During this time the engine was allocated to York, Scarborough, Leeds Holbeck, Royston (at the time of this photograph), Feltham and Eastleigh.

Arnold Road D207
English Electric Type 4, BR Class 40, diesel-electric locomotive D207 was erected at the Vulcan Foundry in July 1958, as part of the first batch of ten engines, and it was allocated to Hornsey shed. The depot utilised the class on services to Sheffield, which could be the duty D207 is pictured here with, on 20 September 1958. The locomotive was condemned after a derailment in February 1983 and the last of the Class 40s were withdrawn in January 1985.

Arnold Road 6925
One of the later members of Charles Collett's GWR 4900 or 'Hall' Class of 4-6-0s, no. 6925 *Hackness Hall* is seen here in a filthy condition on 5 September 1964. The locomotive had worked into the area from its final allocation to Oxley shed, Wolverhampton, and was soon to be withdrawn in November. Two hundred and fifty-nine engines in the class were constructed at Swindon Works between 1928 and 1943 and *Hackness Hall* had entered traffic in August 1941.

Basford North Station

Basford North Station 43155
Departing from Basford North station on 15 March 1958 is Ivatt Class 4MT 2-6-0 no. 43155. The station was opened by the GNR as New Basford on 1 February 1876, but became Basford & Bulwell in August and was later renamed Basford North on 21 September 1953 to differentiate it from the former Midland Railway station. Basford North closed on 7 September 1964 with the cessation of passenger services on the GNR Derbyshire and Staffordshire Extension line.

Basford North Station 61974
Just east of Basford North station and moving some empty carriages to the former GNR carriage sidings is Gresley K3 Class 2-6-0 no. 61974. The engine was built at Darlington in November 1936 and was part of the 1302 series, which featured some alterations from earlier engines. They included; axlebox wedges with angled bolts, stool seats for the enginemen and front footsteps. The track in the foreground from the locomotive is Bagthorpe Curve, which took 'down' traffic from the GCR main line to the GNR line at Basford East Junction.

Basford North Station 61745

After departing from Basford North station during July 1957 Gresley GNR H3 Class, LNER K2 Class, 2-6-0 no. 61754 has been photographed to the east and just past the GCR main line. The H3 Class were a development of the H2 Class, which had been designed for freight traffic, and they possessed a larger diameter boiler, 5ft. 6in., from the latter's 4ft. 8in boiler. The opportunity was also taken to fit a 24 element Robinson superheater. No. 61745 was erected at Doncaster in October 1916 and was condemned in November 1960.

Basford North 61974

K3 no. 61974 is seen here on 12 March 1958; two months before the previous picture of the engine was taken. When built the locomotive was also fitted with vacuum brakes, which had become the standard for new members of the class from 1934, after a number of variations existed previously. Two reservoirs were also originally placed at the rear of the tender, but this was reduced to one on the right-hand side from 1936. No. 61974 was withdrawn from Immingham in July 1962; in both pictures the locomotive's allocation was to Colwick (February 1955-December 1960).

Basford Sidings

Basford North Sidings 69615

A. J. Hill introduced the Great Eastern Railway L77 Class 0-6-2T locomotives in 1915 for London suburban services and twelve were built before Grouping. The LNER was suitably impressed with the design to produce a further ten between 1923-1924 (no. 69615 entered traffic from Stratford in January 1924) before using it as the basis of a Group Standard design. No. 69615 is seen at Basford North sidings on 10 March 1953 and was allocated to Colwick at the time, having arrived in May 1951. The locomotive left for Stratford in April 1954 and was withdrawn in September 1960.

Basford Tunnel

Basford Tunnel 61223

Thompson B1 no. 61223 is steaming into Basford Tunnel, which took traffic from the former GNR line on to the GCR main line into Nottingham and was known locally as the 'Rat Hole'. The tunnel was 90 yards long and passed under the GNR line before joining the GC at Bagthorpe Junction. No. 61223 was built by the NBLC's Queens Park Works during August 1947 and began its service life at Leicester. The engine's career came to an end at Immingham in January 1966.

Basford Tunnel 64354

Running light engine into Basford Tunnel is J11 0-6-0 no. 64354. Construction was carried out by Gorton Works in November 1903 and, like its classmate no. 64406 seen on page 18, subsequently underwent a series of modifications that altered the engine from its original condition. No. 64354 was superheated by the LNER in August 1941 and the company later fitted piston valves in September 1943. At some point the locomotive has been fitted with a plain chimney and small dome to bring it into the Loading Gauge.

Bestwood Park Junction

Bestwood Park Junction 78013

Bestwood Colliery was sunk in 1871 and on 28 July 1873 an Act was passed for the MR to construct Bestwood Park Branch to it and this was completed in April 1874. The line left the MR's Leen Valley line between Nottingham and Mansfield at Bestwood Park Junction, which is the location for this picture of BR Standard Class 2 no. 78013. Erected at Darlington Works in January 1954 the locomotive was allocated to Kirkby-in-Ashfield when working this local service on 13 May 1962. The engine left service from Bolton in May 1967; Bestwood Colliery also closed in 1967.

Bulwell Common

Bulwell Common 45643
Stanier Jubilee 4-6-0 no. 45643, built at Crewe Works in December 1934, was later named *Rodney* at the end of October 1937. The locomotive was one of fifteen engines with straight throatplate fireboxes that were transferred to England from Scotland in 1952 to standardise the type of firebox that was used by the class in the latter country. Fourteen Jubilee locomotives with a sloping firebox throatplate replaced them as well as a rebuilt Fowler 5XP 'Patriot' Class 4-6-0. No. 45643 is seen on 5 August 1964 passing through Bulwell Common on the former GCR line.

Bulwell Common 48212
LMS Stanier 8F Class 2-8-0 no. 48212 was constructed by the NBLC during August 1942. The locomotive was allocated to Colwick shed when it was pictured at Bulwell Common on 10 July 1966 and had been there since the beginning of the year. The abbreviation 'COLK' has been painted on the smokebox door instead of the shed plate, which was a feature on a number of 8Fs based at Colwick towards the end of steam. No. 48212 was removed from service while at Patricroft in June 1968.

Bulwell Common 48696

Also working through Bulwell Common on 10 July 1966 was another 8F, no. 48696, constructed at Brighton Works in April 1944. The locomotive's initial tender pairing was with no. 10404 that had a 4,000-gallon water capacity and a welded water tank. This tender would have also been equipped with pipes mounted inside the coal space to spray water on the coal to keep the dust down and it took its supply from the exhaust injector delivery pipe. No. 48696 ceased to work from December 1967.

Bulwell Common 48699

The majority of the 8F Class were constructed with built-up crescent balance weights set at 50 per cent of the reciprocating masses. During the Second World War the engines built for the Railway Executive Committee had wheels with integrally cast balance weights that were not balanced due to wartime metal shortage. No. 48699 was fitted with this type of wheel when entering traffic from Brighton Works in May 1944. Wheelsets were interchangeable and no. 48699 has since received a set of wheels that were balanced for 50 per cent of the reciprocating masses, which is indicated by the star on the cab side.

Bulwell Common 63644

In its rebuilt form at Bulwell Common station sidings during the mid-1960s is O4/8 Class 2-8-0 locomotive no. 63644. The engine was altered in accordance with a Thompson standardisation programme, which specified diagram 100A boilers used by the LNER B1 Class, but working at 180 psi, should be fitted to locomotives from the O4 Class. New cabs were also added and no. 63644 was dealt with in August 1954; ninety-eight other O4 Class engines were changed to form the O4/8 sub-class.

Bulwell Common 73045

BR Standard Class 5 4-6-0 no. 73045 passes through Bulwell Common station on 31 August 1963 with an express passenger service. The engine was paired with a BR1 tender, which had a water capacity of 4,250 gallons and coal capacity of 7 tons. Originally the BR1 tender dispensed with the traditional fall plate, but after complaints of draughtiness from crews a draught excluder was fitted to eradicate the problem and in later tenders the fall plate was reinstated. No. 73045 was condemned in August 1967.

Bulwell Common 69801

This 4-6-2T locomotive was the second of Robinson's 9N Class to be completed at Gorton Works in April 1911 as GCR no. 166. Twenty-one were in traffic at Grouping and twenty-three were completed for the LNER, who classified them A5. In the 1920s the locomotive was one of several class members to be converted for oil burning. This was in response to a number of coal strikes and this adaptation was present on the engine from June to August 1921 and July 1926 and March 1927. No. 69801 left service in March 1960.

Bulwell Common 63689

The Thompson O1 Class, rebuilt from members of the O4 Class, featured diagram 100A boilers working at 225psi with a 24 element superheater, 20in. diameter by 26in. stroke cylinders with 10in. piston valves and Walschaerts valve gear. No. 63689 was transformed in August 1945 having originally been classified O4/3 when taken into LNER stock during May 1924. R. Stephenson & Co. had erected the engine in March 1918 for the ROD. The locomotive left service in November 1962.

Bulwell Common 90393

War Department 'Austerity' 2-8-0 locomotive, no. 90393 (only just discernable through the obligatory grime), passes through Bulwell Common on 18 April 1964 with a load of coal. The engine had been built at the NBLC's Hyde Park Works during February 1944, as WD no. 8600, and after the locomotive returned from the continent it was stored at Doncaster for a month before moving to Vickers-Armstrongs Scotswood Works in May 1947 for light repairs. After spells at March and Immingham, Colwick was reached in October 1960 and the shed would accommodate the engine until withdrawn on 22 August 1965. Colwick was home to 189 WD Austerity 2-8-0s from the late 1940s for varying amounts of time; five members of the class had a total of four allocations to the shed.

Bulwell Common 61406

The final order for ten Thompson B1 Class locomotives was placed at Darlington Works on 24 February 1949. Nos. 61400-61409 entered traffic in a four-month period, with the first being constructed in March 1950 and the last of the 410-member class ready for service in June; no. 61406 was erected in May. The final ten engines, as well as nos. 61340-61359, were fitted with cast bogie stretcher plates that were slotted and had their size increased from 10½in. deep at the outsides to 1ft. 2½in. Nos. 61400-61409 were also not fitted with electric lighting, but no. 64106 is equipped with longer lower lamp irons. No. 61406's career ended at Doncaster in April 1966.

CHAPTER TWO

Bulwell Hall

Bulwell Hall 61066
Thompson B1 no. 61066 has been pictured on 11 June 1957 running light engine on the former GCR main line at Bulwell Hall. The line passed through the eastern side of the hall's grounds and the latter was soon to be demolished when this image was captured. The land has since become known as Bulwell Hall Park, being used for recreational purposes. No. 61066 was built by the NBLC's Queens Park Works in August 1946 and was withdrawn in September 1962, later being sold to the Central Wagon Co., Ince, for scrap.

Bulwell Hall 44912

The construction of no. 44912 marked the return of Monel metal firebox stays after they had been replaced with steel and copper stays for a period because of wartime metal shortage. Also, the engine was initially paired with tender no. 10537, which was the first to have its tank constructed using both rivets and welding. No. 44912 entered traffic in November 1945 and was in service for twenty-two years, being withdrawn in September 1967. The image dates from 1957 – note the reversed headboard.

Bulwell Hall 61420

At the head of a through freight on 11 June 1957 is North Eastern Railway Vincent Raven Class S3, LNER B16, 4-6-0 locomotive no. 61420. Both Gresley and Thompson rebuilt members of the class, with the resultant locomotives classified B16/2 and B16/3 respectively. No. 61420 was modified to the latter in October 1945 and, amongst other alterations, was fitted with; Walschaerts valve gear, Diagram 49A boiler, left-hand drive and higher running plate. The engine was a long-term York North shed resident, but left service from Hull Dairycoates in September 1963.

Bulwell Hall 61150

Sheffield-based B1 no. 61150 hauls a local passenger service through Bulwell Hall during mid-1957. The locomotive would have been close to undergoing a General repair at Darlington Works between September and October, where it would receive a fresh boiler, no. 28820, which had come from no. 61154 – also allocated at Sheffield. Both locomotives were a product of Vulcan Foundry with no. 61150 being completed in April 1947. The engine was condemned at Sheffield in September 1962.

Bulwell Hall 61839

No. 61839 was erected at Darlington Works in January 1925 and later became the first K3 to be adjusted to have long travel valve gear in October 1928 as LNER no. 134. The original setting had the maximum cut-off at 75%, maximum valve travel at 6⅜in., steam lap at 1½in. and steam lead was 1½in. The new setting was; maximum cut-off 65%, maximum valve travel 5⅝in., steam lap 1⅝in. and steam lead ⅛ of an inch. During a subsequent trial, testing the new arrangement against the old, the saving in coal amounted to 7½ lbs per mile. No. 61839 left traffic in January 1962.

Bulwell Hall 61975
Two Gresley K3s head a freight train on 11 June 1957. The leading locomotive is no. 61975, which was built by Darlington Works in November 1936 as LNER no. 3815. The second locomotive is unidentified, but it is one of the K3s built before April 1925 as it has right-hand drive. The two variations of Group Standard tenders are also illustrated here; straight sides were introduced in 1929 because the stepped-out sides were displeasing to Gresley's eye. No. 61975 was withdrawn from Low Moor shed in September 1961.

Bulwell Hall 63768
Running light engine at Bulwell Hall in the mid-1950s, perhaps to Annesley shed, which was the locomotives allocation between November 1950 and November 1957, is O1 Class engine no. 63768. The concentration of the class was initially at Gorton shed where forty locomotives were allocated. By 1951 Annesley shed had taken the responsibility for the class and housed 53 of the 58 O1s and they were used primarily on the Annesley to Woodford Halse freight service, proving themselves more capable than the previously employed O4s and WD 2-8-0s.

Bulwell Hall 63795
Another O1 Class 2-8-0 travels past Bulwell Hall during 1955, possibly with a Annesley to Woodford Halse freight train. The service was overhauled by the LNER in 1947 by scheduling water stops for down and up trains and making running times stricter. The use of O1s as the primary engines for the service ceased in 1957 when the BR 9F Class 2-10-0s took the reins and twenty-five O1s were transferred away to March shed. No. 63795 was amongst this number and worked from the shed until December 1959. The engine was withdrawn from Staveley depot in October 1963.

Bulwell Hall 64292
J11 no. 64292 was constructed by Neilson Reid & Co. in December 1901. The locomotive was subsequently superheated by the GCR in October 1916, but the apparatus was lost when a saturated boiler was fitted in March 1922. The LNER returned the apparatus to the engine in October 1931 and at the same time brought it within the Loading Gauge. No. 64292 was at Annesley for Nationalisation and moved to Staveley in August 1955. Reallocation to Sheffield Darnall occurred in May 1960 and the engine left service in July 1962.

Carlton and Netherfield

Carlton and Netherfield 44313

LMS Fowler Class 4 0-6-0 no. 44313 is approaching the station at Carlton from the east on the former Midland Railway line between Nottingham and Lincoln, which had been opened in 1846. The locomotive was built at St Rollox Works in November 1927 and withdrawn in December 1959; no. 44313 was allocated to Nottingham shed for a large portion of its career. Behind the coal wagons and to the left of the locomotive is the former GN line to Colwick sidings and shed.

Carlton and Netherfield 62667

LNER D11/1 Class 4-4-0 locomotive no. 62667 *Somme* was built at Gorton Works in November 1922 as part of a batch of six, which were all named after battles that took place during the First World War. The engine was also one of five that were allocated to Lincoln shed during the mid-1950s, no. 62667's allocation lasting from October 1953 to April 1957, and the class could find employment on passenger services to Derby and Nottingham. *Somme* left service from Sheffield Darnall in August 1960.

Chaddesden Sidings

Chaddesden Sidings 40538
Photographed after it was withdrawn in May 1959, at Chaddesden sidings, Derby, is MR Johnson 483 Class, later LMS Class 2P, 4-4-0 locomotive no. 40538. Erected at Derby Works in September 1899 the engine was also rebuilt there in January 1914. This saw the locomotive receive new cylinders, frames and a G7 superheated boiler as well as a 3,500-gallon tender. The locomotive's last allocation was to Derby shed after relocating from Sheffield Millhouses in October 1957.

Chaddesden Sidings 43584
Another engine stored at Chaddesden sidings is MR Johnson 1798 Class Standard Goods 0-6-0 no. 43584. The locomotive was built for the MR by Kitson & Co. during October 1899 and its career lasted until June 1959. Chaddesden sidings were located to the east of Derby station and on the north side of the former Midland Counties Railway line to Nottingham. Installation of the sidings began with extensive earthworks in the early 1860s and by the mid-1870s approx. £30,000 had been spent on three sets of sidings, wagon repair shop and sidings and a twelve-track carriage shed.

Chaddesden Sidings 41795

Chaddesden sidings increased in importance during the mid-1950s as a number of the smaller yards around Derby were closed. However, its role was soon diminished by colliery closures in the area. Some of the free space in the sidings was allocated to locomotives and wagons that had been marked for disposal and one such engine was this MR Johnson 0-6-0T locomotive no. 41795. Much of the remaining freight traffic was subsequently rerouted to Toton and Chaddesden had closed by the mid-1980s. Much of the site is now utilised as an industrial estate.

Colwick Shed

Colwick Shed 61159

Stood in Colwick shed's yard on 19 April 1955 is Immingham-based Thompson B1 no. 61159. The locomotive was built by the Vulcan Foundry in May 1947 and spent the first seven years of its career allocated to Gorton shed before moving to Immingham in June 1954. From the latter, passenger services could be taken to London, Birmingham via Nottingham, and Manchester in addition to local passenger trains. The locomotive left service in September 1963 and later in the year was sold to J. Cashmore, Great Bridge, for scrap.

Colwick Shed 61367

Another B1 at Colwick, on the same day as no. 61159, was no. 61367, which had been allocated to the shed from Leicester during April 1952. To the left of the locomotive Colwick's 500 ton mechanical coaler is visible and this was in operation at the site by early 1939. It had been installed as part of a number of improvements authorised in 1936, which also included the provision of a wet ash pit, 70ft. turntable, a new well and new water columns; the total cost was approx. £35,000. No. 61367's career ended at Doncaster during August 1965.

Colwick Shed 63675

O4/8 locomotive no. 63675 undergoes maintenance in Colwick yard, with some of the motion removed and placed on the ground beside the engine. No. 63675 was constructed by Robert Stephenson & Co. Ltd in October 1917 for the ROD and was later bought by the LNER and taken into traffic during May 1924. Rebuilding to O4/7 occurred in December 1939, while further alterations to O4/8 took place in February 1957. The locomotive spent its last ten years in service working from Colwick and was condemned in January 1966.

Colwick Shed 64273

At the ash pits is former GNR J22 Class 0-6-0, later LNER Class J6, locomotive no. 64273. Fifteen engines were erected to the design of H. A. Ivatt before he retired and was succeeded by H. N. Gresley, who subsequently made some alterations to the design of the following ninety-five class members. These locomotives were designated the 536 series and had a repositioned boiler, shorter cab, sandboxes above the running plate and a shorter chimney; no. 64273 was constructed at Doncaster in March 1922 towards the end of the series

Colwick Shed 68028

No. 68028 was built by the Hunslet Engine Co. in April 1945 as part of the War Department's 'Austerity' Class of 0-6-0ST locomotives designed by R. A. Riddles. In July 1946 the engine became a part of LNER stock, classified J94, and lost its WD no. 71447, acquiring LNER no. 8028. Immingham was the locomotive's allocation for the majority of the 1950s before a transfer to Colwick transpired in December 1958. No. 68028 was condemned at the shed in September 1960.

Colwick Shed 64802

Colwick repair shop provides the backdrop for this Gresley J39 Class 0-6-0 engine, which was constructed at Darlington Works in October 1929. The two-road repair shop at the shed was erected during the early 1880s and was furnished with £1,215 worth of equipment, allowing a variety of repairs and maintenance to be undertaken on site; facilities were further enhanced by a 55ft. extension in 1897. No. 64802 is on the westernmost road of the original engine shed (known as the 'old shed'), which had been built for the GNR in 1876.

Colwick Shed 67799

To the east of the repair shop, which is just visible to the right, is Thompson L1 Class 2-6-4T engine no. 67799. The engine was amongst the initial L1s to be allocated to Colwick in April 1954 (replacing a number of N7s that were relocated to Stratford, Woodford Halse and Hatfield) and they were generally used on passenger services to Derby and Grantham. A total of forty-two were based at the shed and a high of thirty-one engines was reached in the early 1960s. No. 67799 was withdrawn from the shed in March 1962.

Colwick Shed 68554

No. 68554, a former Great Eastern Railway Class R24 0-6-0T, was built at Stratford Works during October 1895 to the design of J. Holden. The class consisted of engines that were used for passenger and shunting duties. Those used for the former featured; Westinghouse brakes, screw reverse and balanced wheels. Some of the class were rebuilt by the GER (GER R24 Rebuilt) with a larger boiler with working pressure raised to 180 psi and an increased firebox grate area; no. 68554 was changed in July 1904. Allocation to Colwick lasted from March 1958 to October 1959.

Colwick Shed 69360

No. 69360 was three months away from a transfer to Gorton shed when pictured at Colwick on 19 April 1955; it had been allocated to the latter since November 1953. The locomotive had begun its career in June 1900, as GCR no. 936, emerging from Beyer, Peacock & Co. as part of the GCR's 9F Class, which were designed by Thomas Parker. The class were intended for use of short goods trips and shunting, but occasionally were pressed into a passenger duty; the N5s at Colwick were generally employed as shunters. No. 69360 was withdrawn in March 1960.

Colwick Shed 90202

In Colwick shed's yard, with at least two other classmates, is WD Austerity 2-8-0 no. 90202. After being constructed by the NBLC's Hyde Park Works in October 1943, the engine went on loan to the LNER until December 1944, when it was transferred to the Southern Railway. In 1947 the locomotive was again loaned to the LNER and worked from Colwick, Annesley and Woodford Halse before Nationalisation. No. 90202 had a twelve-year stay at Colwick from 1948, broken by two weeks at Staveley in April 1951, and was withdrawn from Canklow in April 1965.

Colwick Shed 90103

Three hundred and fifty Austerity 2-8-0s were loaned to the LNER during the Second World War and no. 90103 was amongst this number, working for the company from being completed in February 1943 until December 1944 when it was prepared for deployment to the continent. The engine was back with the LNER in February 1947 and at Nationalisation a total of 270 of the class were employed by the LNER. No. 90103 was allocated to Colwick from August 1949 until leaving service in November 1965.

Colwick Shed 68950

This image appears to have captured LNER Gresley J50 Class 0-6-0T locomotive no. 68950 in the midst of some repairs close to the shear legs at Colwick shed. The engine was erected at Doncaster Works in August 1926 with Group Standard design features that were also applied to thirty-seven other locomotives constructed between 1926 and 1930. No. 68950 was also built with plain coupling rods, but has since acquired fluted coupling rods. The locomotive was at Colwick between July 1956 and February 1960 and left traffic in September of the following year.

Colwick Shed 61439

B16 Class 4-6-0 no. 61439 rests slightly to the east of the old shed, next to the stores and the shed's shear legs are visible in the background; these were installed during 1924, had a capacity of 35 tons and were originally located at Grimsby. The shear legs were powered by gas drawn from the mains. In *Great Northern Railway Engine Sheds Volume 2* (1996) it is related that when they were in operation they could cause havoc with the gas supply to the houses in the vicinity and many complaints were received when it was lost at meal times!

Colwick Shed 49323

Two 'foreigners' have been caught at Colwick on 18 November 1949; one is former London & North Western Railway G1 Class 0-8-0 no. 49323, while the other is unidentified, but is probably an 8F 2-8-0. The LNWR previously had permission to use the shed until the company erected their own premises when the joint line with the GNR opened in 1879. The LMS closed this 8-track shed in 1932 and the company were subsequently awarded two roads to use at Colwick. No. 49323 was based at Watford in 1949 and after a number of moves was withdrawn in September 1962.

Colwick Shed 61768

In a pre-emptive move against an expected increase in traffic (especially coal, which was the focus of work at Colwick) from the opening of the GNR's Leen Valley line, the company built a new engine shed on the west side of the original one and it was separated by the repair shop – also installed at this time. The shed, 275ft. long (later extended to 330ft.) by 115ft. wide, was completed in mid-1882 and had eight tracks covered by a north light roof. Gresley K2 no. 61768 stands outside the shed ready for work *c.* 1955.

Colwick Shed 63721

O4 2-8-0 no. 63721 had been a recent visitor to Gorton Works when this picture was taken in late 1949. At a General repair, carried out in August, the locomotive received a fresh boiler and its BR number. No. 63721 had been constructed by Robert Stephenson & Co. in June 1919 for the ROD and was on loan to the LNWR for a time before being stored by the government. As ROD no. 1645, the locomotive was taken from Queensferry dump and entered traffic for the LNER, as no. 6331, in June 1924. The engine was rebuilt as O4/8 in April 1954 and was withdrawn in November 1962 from Gorton.

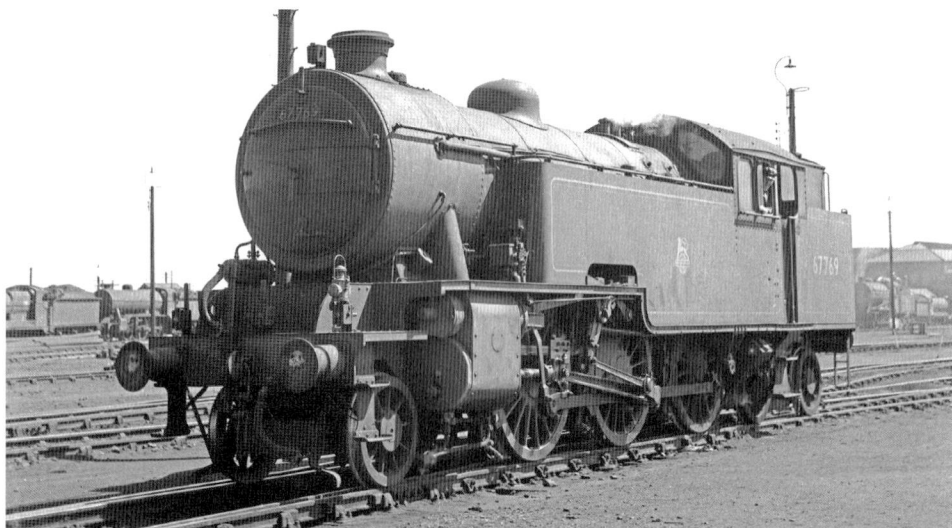

Colwick Shed 67769

The L1 Class were plagued by hot axleboxes and a number of attempts were made at solving this problem. When a reduction in the number of cases was reached, the class were being withdrawn and application to further locomotives was deemed uneconomical. No. 67769 was part of a trial, which saw the engine fitted with manganese steel axlebox liners with horn wedges removed; nine other based at Neasden shed underwent the alteration during late 1954. The locomotive had just been reallocated from Neasden when it was pictured at Colwick in late April 1955.

Colwick Shed 68873

Parked on a track that ran adjacent to the wagon shop at Colwick is former GNR Ivatt Class J13 0-6-0ST locomotive, LNER Class J52, no. 68873, erected at Doncaster Works in October 1905. The class were a development of Patrick Stirling's GNR J14 Class and they had a 4ft. 5in. diameter domed boiler; this being a 4½in. increase over the J14s. The locomotive arrived at Colwick from King's Cross shed during July 1952 and was withdrawn from the former in September 1955. Note that one of the stationary boilers used by the wagon shop is visible in the background.

Colwick Shed 69064

On the east side of the old shed in late 1949/early 1950 is GCR Robinson 1B Class 2-6-4T no. 69064. The track the locomotive is resting on runs over the site of the original 45ft. turntable (the outline just discernible) and behind no. 69064 is the shed's water tank. Water was initially collected from a weir in Carlton Beck, but after causing a significant flood in the early 1880s, the GNR were forced to sink a borehole on the east side of the old shed. Water was reached at a depth of a depth of 181ft. and was deemed to be of high quality, with further improvement being seen after softening treatment.

Colwick Shed 69822

Open to the elements in the early 1950s before being re-roofed, Colwick shed provides little cover for this A5 4-6-2T, no. 69822. The locomotive was built at Gorton in February 1923 with a side window cab and one-piece front spectacle plate; the latter had been made of two pieces for previously constructed engines. Two modifications implemented by the LNER and visible here are the strengthening stay attached the front of the water tank, which curved under the boiler to the opposite side, and the two handles that replaced the original wheel and handle used for opening the smokebox. No. 69822 left service in November 1958.

Colwick Shed 90221

The WD Austerity 2-8-0s featured; 4ft. 8½in. diameter driving wheels, 3ft. 2in. diameter leading wheels, a 5ft. 8½in. diameter boiler working at 225psi, two outside cylinders 19in. diameter by 28in. stoke and a 28 element superheater. No. 90221 was constructed by the NBLC's Hyde Park Works during April 1943 and was another WD engine to be loaned to the LNER. No. 90221 has been photographed at Colwick on 15 April 1958 and is visiting from Immingham; it left service from there a while later in January 1965.

Colwick Shed 2173

This former GNR 4-4-0 was pictured on 13 November 1949 and only had seven months of its career remaining before it was condemned for scrap in May 1950. No. 2173 began life at Doncaster in November 1900 as part of the D1 Class (LNER D2) designed by Ivatt. At Nationalisation the engine was one of three allocated to Grantham, while thirteen of the class could be found at Colwick; the last D2 was withdrawn from the latter in June 1951. Colwick shed closed to steam on 12 December 1966, but had been gradually stripped of its importance since the late 1950s. The shed was closed fully on 13 April 1970 and the site was later cleared.

Daybrook Station

Daybrook Station 61914

Waiting at the 'up' platform for the signal to proceed with the 1.32 p.m. Basford North to Nottingham Victoria station via Gedling on 15 March 1958 is Gresley K3 no. 61914. The locomotive was one of twenty built by Armstrong Whitworth & Co. and entered traffic in May 1931; these locomotives were classified K3/5 until December 1935 (becoming K3/2) as they had no axlebox wedges and were fitted with solid hornblocks instead. No. 61914 would move to Immingham in August 1960 before returning to Colwick in September 1961 and the engine was withdrawn in August 1962.

Daybrook Station 63589

Daybrook station, located between Basford North station and Arno Vale station on the GNR's Derbyshire & Staffordshire Extension line, opened on 1 February 1876 as Bestwood & Arnold station. The change of name occurred shortly later in March, but Quick (2009) notes that various combinations of Daybrook, Bestwood and Arnold were used before Daybrook became the exclusive name after 1947. This O1 Class locomotive, constructed as 8K by the NBLC's Hyde Park Works in December 1912, is heading a mineral train through the station during the late 1950s.

Daybrook Station 61896

This scene, captured from the west end of the 'down' platform, sees Gresley K3 no. 61896 switch from the 'up' goods line to the 'up' main line. To the right is Daybrook station's goods shed, which had two tracks to the right of the building for coal wagons, weighbridge, offices, dock and a 10-ton crane. No. 61896 was erected at Darlington Works in August 1930 and was in service until May 1962; the final eight years of its career was spent working from Colwick. Daybrook station closed on 4 April 1960, but goods services ran until 1 June 1964; the track was lifted during 1966.

CHAPTER 3

Derby

Derby Midland Station 45543

Departing from the south end of Derby Midland station with the Locomotive Club of Great Britain's 'The Midland Limited' railtour on 14 October 1962 is LMS Fowler 5XP 'Patriot' 4-6-0 no. 45543 *Home Guard*. The engine had been attached to the train at the station and would guide the tour to Loughborough, Leicester, Market Harborough, Kelmarsh and Northampton before Stanier Class 5 no. 45392 would take the tour to its conclusion at London's St Pancras station. No. 45543 had come from Carnforth for the railtour and had been at the shed since June, but the allocation would not last any longer than November as the locomotive would be condemned for scrap.

Derby Midland Station 40935

In the early 1950s a renewal scheme began at Derby Midland station, which saw the old three-bay train shed removed and replaced by a canopy on each platform. The work had partially been brought about by damage sustained during an air raid on 15 January 1941 when 300ft. of the train shed roof was destroyed, in addition to part of platform six. The construction work was completed in mid-1954 and had cost approx. £200,000. LMS Fowler Class 4P Compound 4-4-0 has been photographed at the station on 8 March 1953, with part of the old train shed still standing and visible in the background.

Derby Midland Station 44113

The diagonal yellow stripe on the cab side was applied to a number of LMS Fowler 4F Class 0-6-0s from September 1964 when the restrictions under the electric overhead lines south of Crewe came into effect for locomotives over 13ft. 1in. high.. However, a number of the class were under the limit and would have been free to use the line but for confusion over which engines were eligible. No. 44113 was one of twelve LMS-built 4Fs to survive into 1966, but was one of three class withdrawals during January. The locomotive is pictured leaving Derby Midland station during 1965.

Derby Midland Station 45590

No. 45590 *Travancore* has stopped at Derby Midland station – note the new platform canopies – with the York to Bristol express passenger service on 12 May 1956. The locomotive was allocated to Sheffield Millhouses shed, which usually supplied the engine for the Sheffield to Bristol portion of the service and the return journey was also likely to be undertaken. No. 45590 had been at Millhouses since April 1947 and, apart from two loans in 1951, would be at the shed until December 1961. *Travancore* subsequently left service in December 1965 from Warrington.

Derby Midland Station 60019

Making an extremely rare appearance at Derby Midland station on 6 March 1966, with the Williams Deacon's Bank Club railtour, is Gresley A4 Class Pacific locomotive no. 60019 *Bittern*. The outing ran from Manchester Piccadilly station to Crewe and then Derby with visits to the latter two places' locomotive works also part of the trip.

Derby Midland Station 48618

Toton-based Stanier 8F 2-8-0 no. 48618 stands at the south end of Derby Midland station on 24 March 1956 and judging from its appearance has just received attention at Derby Works. Along with Crewe and Horwich, Derby was responsible for maintenance of the 8F Class, with Bow and Rugby also receiving engines in need of light attention. No. 48618 had been constructed by Ashford Works in September 1943 and its career lasted for twenty-four years, being withdrawn from Lostock Hall in September 1967.

Derby 48543

Erected at Darlington in January 1945, 8F no. 48543 was part of an order for twenty 8F engines placed at the works in January 1943 by the Railway Executive Committee. A total of sixty were built for the REC at the LNER works at Doncaster and Darlington and were taken on loan until they could be transferred back to the LMS; no. 48543 was returned in March 1947. A feature of the locomotives built by the LNER were riveted tender tanks and disc wheels, with the latter still in use on no. 48543's tender. The engine completed its service life in February 1966.

Derby Shed 41748

MR Johnson 1377 Class 0-6-0T locomotive no. 41748 was constructed at Derby Works in the latter half of 1884. The engine is stood to the side of Derby no. 4 shed in 1951 when it was a resident at Burton shed; the locomotive had seen in Nationalisation at Skipton. No. 41748 would move to Gloucester Barnwood shed in March 1952 and would be withdrawn from there in September 1957 – it was just under seventy-three years old.

Derby Works 41516

The 0-4-0ST locomotives built by the MR to Johnson's design were unclassified by the company, but the LMS classified the engines 0F. This example had been erected at Derby in the final years of the nineteenth century as part of order no. 1534 and these locomotives had slightly increased weight over the 0-4-0STs produced pre-1893. No. 41516 was one of only four of the type to survive into BR service and was the penultimate engine to be withdrawn in October 1955. No. 41516 is seen at Derby in mid-1955.

Derby Shed 41724

Derby's no. 4 shed was installed in 1890 by Messrs. Walkerdine for a price of approx. £15,000. The shed, a double roundhouse, was located to the south of Derby Midland station and on the east side of the lines emanating from that end. Stood on track running the length of the shed is MR Johnson 1377 Class 0-6-0T locomotive no. 41724, built in the early 1880s at Derby, and classmate no. 41773 is also present. The former worked from Derby between May 1955 and June 1958 when it left service.

Derby Works 41192

Minus its tender at Derby Works on 14 May 1956 is LMS Fowler Class 4 Compound 4-4-0 no. 41192. The class were a development of the MR 4-4-0 Compounds and had a superheated boiler and slightly smaller driving wheels than their pre-Grouping counterparts. No. 41192 was built at the Vulcan Foundry during March 1927. The locomotive had been allocated to Derby since August 1952 after spells at Blackpool and Kettering since 1948. The engine was condemned in June 1957.

Derby 40059

The LMS Fowler 3P Class 2-6-2T locomotives were introduced in 1930 for light passenger services. No. 40059, erected at Derby during October 1931, displays some of the alterations, which include; outside steam pipes, greater diameter chimney and vacuum controlled regulator valve. The engine has also been fitted with push and pull apparatus. No. 40059 was based at Hull Dairycoates when pictured at Derby during April 1955, but would leave service from Heaton Mersey in November 1959.

Derby Shed 40513

MR Johnson Class 483 4-4-0 was constructed by Sharp, Stewart & Co. in December 1899, but was rebuilt by the MR in November 1912. From 1919 a scheme was started to modify the tenders paired with the 483 Class. However, this was discontinued in 1930 leaving a number unaltered and no. 40513 has a MR Johnson 3,250 gallon tender that has not been changed. The locomotive also has an exhaust steam injector, visible coming out beneath the smokebox, which was fitted between 1932 and 1937. No. 40513 is in Derby shed's south yard with one of the turntable tracks in the foreground of the picture.

Derby Shed 44806

LMS Stanier Class 5 no. 44806 was erected at Derby Works in July 1944 and was withdrawn in August 1968; the locomotive still had a number of years service left when it was photographed at Derby on 12 June 1955. Subsequently, no. 44806 was bought by Kenneth Aldcroft and had a number of brief spells on preservation lines before a cracked firebox sidelined the engine in 1974. After spending time as a static exhibit, no. 44806 was restored at the Llangollen Railway and became operational in 1995. At the time of writing the locomotive is still at work on the railway.

Derby Shed 44814

A 60ft. turntable, with sixteen radiating tracks, was installed at Derby shed in 1900 and was bought from Eastwood Singler & Co. for £1,065. The shed later acquired a 70 foot turntable and this was removed during the mid-1960s in preparation for the changeover to housing diesel locomotives. No. 44814 was another Class 5 built at Derby and entered traffic in October 1944; fifty-four class members were produced at the works between April 1943 and December 1944. The locomotive was visiting Derby from Saltley shed when pictured on 13 May 1956 and would be withdrawn from Shrewsbury in September 1967.

Derby Shed 45509

Also stood on a road around the turntable at Derby is LMS Fowler Class 5XP 'Patriot' 4-6-0 locomotive no. 45509 *The Derbyshire Yeomanry*, which has been photographed on 24 March 1956. The engine was constructed at Crewe Works during August 1932 and was in service until 1951 before it was named at platform one at Derby Midland station. No. 45509 was allocated to Derby from October 1951 until a move to Newton Heath occurred in August 1958. The locomotive was based at the shed for three years before leaving service in August 1961.

Derby Shed 45626

This view, taken from the turntable at Derby during July 1955, shows LMS Stanier Jubilee 4-6-0 no. 45626 *Seychelles*, LMS Class 4P Compound 4-4-0 no. 41094 and, slightly out of view, Thompson B1 4-6-0 no. 61353. No. 45626 was built by Crewe Works in November 1934 and was in service until October 1965. The engine was in the midst of its second spell at Derby, which lasted between June 1951 and December 1961. No. 61353 was a long way from 'home ground' at this time as it was allocated to Darlington shed. The former GNR shed at Derby was closed in 1955 and could partially explain its appearance here.

Derby Works 21

Perhaps awaiting its new British Railways identity at Derby Works in the early 1950s is LMS Fowler 3P Class 2-6-2T engine no. 21. The locomotive was erected at Derby Works in December 1930 as LMS no. 15521, however in 1934 the class were renumbered from 15500-15569 to 1-70. No. 21 was one of twenty 3Ps to be fitted with condensing apparatus to allow them to work suburban services over the Metropolitan line in London; it has since been removed from the engine. No. 21 ceased to be in service from September 1959.

Derby Works 22851

MR Matthew Kirtley 0-6-0 goods locomotive, LMS no. 22851, stands on the scrap line at Derby Works on 9 September 1949. The design was introduced in 1863 and featured curved frames, which became the standard arrangement until Kirtley was replaced by Samuel Johnson as Locomotive Superintendent in July 1873. Production of the class ceased in 1874 and some members were later rebuilt by Johnson and his successor Richard Deeley. Four hundred and forty-nine were in service at Grouping, but by Nationalisation this had dropped to four and the last of the class, LMS no. 22630, BR no. 58100, was withdrawn in November 1951.

Derby 47324

After Grouping the LMS found that they were in need of a dedicated shunting and goods engine in the Class 3 power category. The MR's 1900 Class 0-6-0T, which had relatively recently been rebuilt with a Belpaire firebox, was chosen to be the standard design. Design work for the new class was carried out by Derby's drawing office, but construction of the first engine was completed at Vulcan Foundry in July 1924. No. 47324 was built by the NBLC's Hyde Park Works in June 1926 and was in traffic until December 1966. The locomotive is one of nine class members that have been preserved and is currently based at the East Lancashire Railway.

Derby Shed 47572

After large orders for the 3F Class had appeared in the 1926 and 1927 locomotive renewal programmes (150 and 100 respectively), an order for a further 100 engines, as part of the 1928 programme, was deemed necessary. As no. 16655, this locomotive was one of twenty-five erected by the Hunslet Engine Co. and it entered service in late September 1928. No. 47572 (from 15 May 1948) is in Derby shed's south yard during July 1955; the shed's water tank is seen behind the engine. The locomotive was withdrawn during May 1962.

Derby Works 10001
In the mid-1940s a section of Derby Works paint shop, 450ft. by 45ft., was partitioned off to become no. 10A diesel shop. This was the birthplace of the two pioneer diesel-electric locomotives, nos. 10000 and 10001, built in December 1947 and July 1948 by the LMS in partnership with English Electric. No. 10001 is pictured in no. 10A diesel shop during July 1955.

Derby 10100
This unusual-looking 4-8-4 diesel-mechanical locomotive was completed at Derby Works in January 1952 and was the result of a partnership between Fell Developments Ltd and BR. The locomotive used four Paxman 500 hp supercharged diesel engines, split evenly between the two ends and two 150 hp engines that were used to power auxiliary components. Drive was delivered to the inner pairs of wheels via a mechanical transmission. The locomotive was in traffic until November 1958 when the gearbox was destroyed by a faulty component and no. 10100 was later scrapped at Derby.

Derby 10201

D16/2 Class diesel-electric locomotive no. 10201 was one of three locomotives built to Oliver Bulleid's design by the former Southern Railway works at Ashford (2) and Brighton (1) between 1950 and 1954. No. 10201 was the first to be completed in December 1950 with a 1,750 hp English Electric diesel engine and six English Electric traction motors. The locomotive was used from depots on the Southern Region, before moving to Camden in 1955 and working on the London Midland region. Withdrawal from Willesden depot occurred in December 1963. No. 10201 has been pictured at Derby on 12 June 1955.

Derby 12025

Derby Works ventured into diesel locomotive design during the early 1930s and produced prototype diesel-hydraulic locomotive no. 1831, which was converted from a Johnson 0-6-0T steam engine. This design was unsuccessful, but after a period of development, the LMS produced another 0-6-0 with an English Electric engine and electric transmission. A total of forty were produced, nos. 7080-7119, and all were built at Derby works; no. 12025 being constructed in February 1942 as LMS no. 7112. Photographed at Derby in July 1955, the locomotive was in service for a further twelve years before being condemned for scrap in November 1967.

Derby Shed 45602
Marked 'Not to be Moved' due to work on the right-hand cylinder, Stanier Jubilee 4-6-0 no. 45602 *British Honduras* stands, in otherwise excellent condition, at the south end of Derby shed on 14 May 1956. The locomotive was on loan to Derby from Bristol Barrow Road between April and June 1956 and would return to the former permanently in September for a year-long allocation.

Derby Shed 46443
Ivatt Class 2MT 2-6-0 no. 46443 emerged from Crewe Works in February 1950 and was allocated to Derby for the remainder of the decade; it is pictured on shed in mid-January 1959. The engine left service from Newton Heath in March 1967 and has been preserved, working on the Severn Valley Railway.

Derby 44070

Fowler 4F no. 44070 was constructed at the NBLC's Hyde Park Works in November 1925. The locomotive was one of twenty-five built there in the latter half of 1925 and a feature that first appeared on this batch was a manual overflow cock on the right-hand injector and a non-return valve on the left-hand side. Previously, for engines constructed for the LMS and MR, a manual overflow cock had been provided for both injectors. However, non-return valves were later fitted on both sides and the manual control was discontinued from LMS no. 4407 onwards. No. 44070 was withdrawn in June 1962.

Derby Shed 44552

No. 44552 is pictured at the north-west end of Derby shed in April 1955 and the shed's offices are seen behind the engine. They survived the shed, which was demolished in 1969 and lasted into the 1970s. The locomotive was erected at Crewe Works during December 1928 as part of an order for fifty engines and these were fitted with carriage warming apparatus at the tender end. Towards the end of the 1920s further 4Fs were given the equipment as it was deemed desirable for the class to have it for use in the winter months. No. 44552 was condemned in September 1964.

Derby 53810

The Somerset & Dorset Joint Railway was leased to the MR and London & South Western Railway in 1876 and, for the MR, the line came with the responsibility of providing locomotives and rolling stock. Due to the nature of the line powerful locomotives (especially for mineral traffic) were required and Fowler entrusted the job of designing a suitable 2-8-0 to James Clayton. The result was the 5P5G, later 7F Class, which initially numbered six when introduced in 1914. The engines had; 21in. by 28in. cylinders, Walschaerts valve gear, 4ft. 7½in. driving wheels and a G9AS boiler. A further five appeared in 1925 with a larger G9BS boiler (later replaced by the G9AS boiler) and this locomotive, as S&DJR no. 90, was the last to be constructed by Robert Stephenson & Hawthorns in August. The engine gave the line 38 years service as was relieved of its duties in December 1963; two of the class have been preserved.

Derby Works 58087

MR Johnson 2228 Class 0-4-4T locomotive no. 58087 was manufactured by Dübs & Co. in August 1900. The engine was withdrawn from Plaistow in July 1955 and is likely to be at Derby during the same month for scrapping.

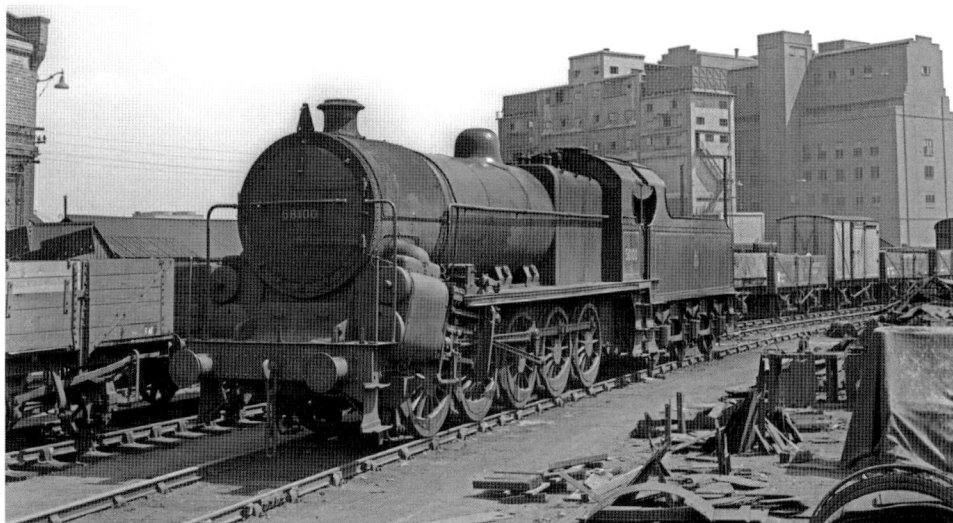

Derby Works 58100

The two mile-long Lickey incline, with a gradient of 1 in 37½ between Bromsgrove and Blackwall on the Birmingham-Bristol line, had provided an expensive problem to the MR for a number of years. In the early 20th century a powerful 0-10-0 locomotive was designed specifically to work the line. MR no. 2290, also known as the Lickey Banking Engine or 'Big Bertha', was built in December 1919 and had; four 16¾in. by 28in. cylinders with 10in. piston valves, Walschaerts valve gear, 4ft. 7½in. diameter driving wheels and a G10S boiler. Having been withdrawn in mid-May 1956 from Bromsgrove shed, no. 58100 has arrived at Derby Works a short time later to be scrapped. The locomotive was replaced by 9F Class 2-10-0 no. 92079, which also acquired the distinctive headlight used by 'Big Bertha'.

Derby Shed 58122

This locomotive was one of fifty 0-6-0s that were ordered as the result of an assessment complied by Johnson when he became Locomotive Superintendent in 1873 of the future traffic requirements. Kitson & Co. were contracted to build twenty 1142 Class locomotives and no. 58122, as MR no. 1157 was erected during August 1875. The engine was in service for a commendable eighty-six years and was withdrawn from Bescot in September 1961.

Derby Works 41938

The London, Tilbury & Southend Railway's 79 Class 4-4-2T engines were perpetuated by the LMS after Grouping in small batches between 1923 and 1930. Thirty-five in total were constructed (classified 3P by the LMS) and this example was erected by Nasmyth Wilson & Co. in August 1925, being one of five built by the company; the remainder were completed at Derby. Almost exclusively a Leicester engine from Nationalisation, no. 41938 left service in February 1955 and Bill has caught the locomotive in the midst of its demise at Derby Works in July 1955.

Derby 42050

Charles Fairburn's only design of locomotive during his brief tenure as CME of the LMS was for a 2-6-4T engine, which was a development of the Stanier 2-6-4T. Between 1945 and 1951 277 examples of Fairburn's design were produced at Derby and Brighton Works; this locomotive was erected at Derby in September 1950 and is seen at the works on 9 June 1951. No. 42050 was allocated to Bourneville (from new), Derby, Trafford Park, Brunswick before returning to Trafford Park to be condemned at the shed in April 1965.

Derby 40933

After Grouping 195 compound 4-4-0 locomotives were built for the LMS, which added considerably to the forty-five that had been inherited from the MR in 1923. No. 40933 was built by Vulcan Foundry in June 1927 as construction of the class was winding down; the works contributed seventy-five engines to the class total. No. 40933 is pictured at Derby Works on 20 November 1955 with a prototype Stanier high-top tender (no. 3677), which had begun life as a Fowler 3,500-gallon tender coupled with a 4F 0-6-0. It was rebuilt in 1933 and paired with 4P Compound, LMS no. 936, until 1954. No. 40933 was withdrawn from Gloucester Barnwood shed in April 1958.

Derby Shed 41906

Before Derby Works was thrown into the production of Stanier's new locomotives, ten updated Johnson 0-4-4T engines were constructed at the works between 10 December 1932 and 14 January 1933. The locomotives had slightly larger driving wheels and cylinders from the earlier engines at 5ft. 7in. and 18in. by 26in. respectively. No. 41906 is seen here at Derby on 21 June 1957 while briefly allocated to Crewe. The locomotive left service from Buxton in November 1959 and all except no. 41900 were withdrawn during the month; the latter left service in March 1962.

Derby 42371

All of the Fowler Class 4P 2-6-4T locomotives were built at Derby Works and no. 42371 entered traffic in September 1929. The class had two 19in. by 26in. cast-iron cylinders with 9in. piston valves, which were similar to those fitted to Fowler's Royal Scot 4-6-0. During the 1940s the cylinders were replaced by ones made from cast-steel and no. 42371 received the new type in October 1948. The smokebox had also been redesigned in the 1940s because of problems caused by the layout of the steam and exhaust pipes as well as corrosion issues. The new smokebox had outside steam pipes and a new blastpipe and exhaust. Withdrawal of the locomotive occurred in April 1962.

Derby Works 42523

In addition to designing a two cylinder 2-6-4T, Stanier produced a three cylinder version in 1934 for working the line to Southend and to replace a number of former LT&SR locomotives. No. 42523 was built at Derby Works during September 1934 with; 16in. by 26in. cylinders, domeless and tapered 4C boiler working at 200psi and 5ft. 9in. diameter driving wheels. From Nationalisation to being condemned in June 1962, the locomotive was allocated to Shoeburyness shed.

Derby Shed 42767
LMS Hughes 2-6-0 no. 13067 was erected at Crewe Works in July 1927 as part of the first lot to be built at the works; eventually 175 members of the class were completed there and 70 engines were the product of Horwich. In the 1934 renumbering the locomotive became no. 2767 in August and later, as part of the BR number alterations was given the 'M' prefix to denote that it was formerly of the LMS. Horwich Works applied the BR number in November 1949.

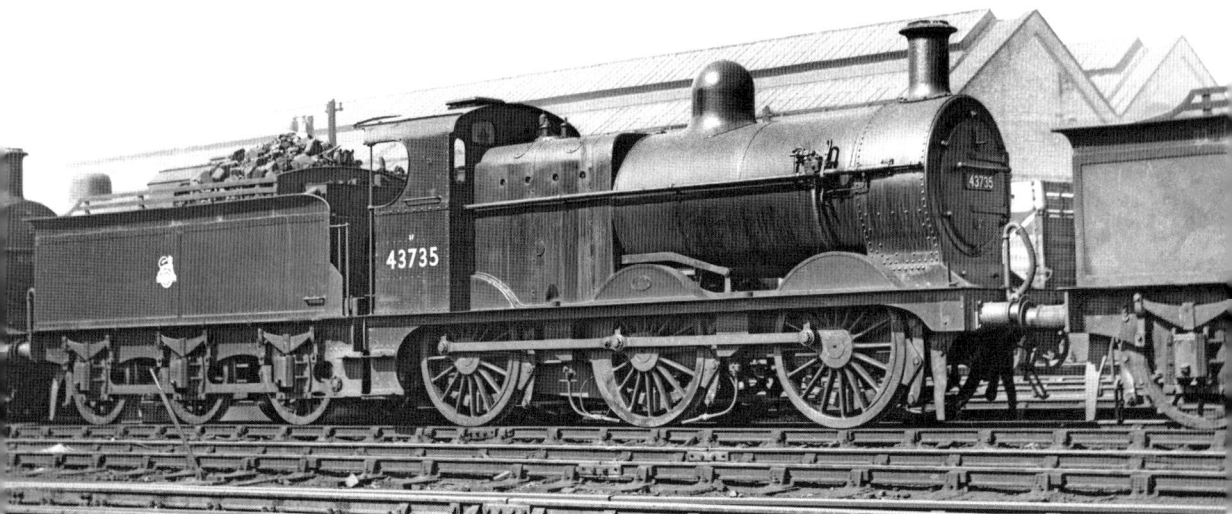

Derby 43735
Entering traffic from Neilson & Co. in January 1902 as MR no. 2706 this locomotive was a late example of Johnson's 1798 or 'Standard Goods' Class. Production of the class began in September 1888 and ceased, shortly after the completion of no. 2706, in August 1902 and there were then 575 engines in service. No. 2706, no. 3735 from May 1907, was subsequently rebuilt with a H boiler in June 1912 and again in May 1923, but with a G7 boiler and new cab, which altered the engine to its appearance here. The engine's time in service came to an end in September 1960 after spending the BR period working from Derby shed.

Derby 73062

BR Standard Class Five 4-6-0 no. 73062 looks truly resplendent after it had been completed at Derby works in September 1954. The engine would soon make its way to Polmadie shed, Glasgow, to begin its career; nos. 73055-73064 built at Derby were all initially allocated to the Scottish Region at the shed. No. 73062 was transferred to Motherwell in October 1955, but returned to Polmadie in June 1957 and was condemned there in June 1965. Derby was responsible for the maintenance of the class allocated to the London Midland Region, while the engines working in Scotland were maintained by St Rollox from 1953 and Cowlairs from 1958.

Derby 75011

Despite resembling Standard Class 5 no. 73062, seen above, this BR Standard Class Four 4-6-0 was more closely related to the LMS Class Four 2-6-4T engines and their development in the Standard series. In comparison with the latter, the 4-6-0 had a greater total heating surface and a slightly lower tractive effort, but they shared the same cylinder and coupled wheel dimensions. No. 75011 was erected at Swindon in November 1951 and allocated to Patricroft. However, when photographed at Derby during April 1955, the engine had been working from Llandudno Junction for the previous eighteen months. The locomotive left traffic in November 1966.

Derby Works 47000

No. 47000 was one of five 0-4-0ST locomotives constructed by Kitson & Co. in 1932 for use on LMS shunting duties; no. 47000 was completed in November. A further five engines were erected to a similar, but modified design, by BR at Horwich in 1955. Based at Burton upon Nationalisation, the engine moved to Rowsley in April 1952 and found its way to Derby in January 1959. Westhouses was fleetingly the locomotive's allocation in May 1960 before it returned to Derby in July. Withdrawal occurred in October 1966.

Derby Works E.D. 1

This 0-4-0 diesel-mechanical shunter was erected by the Leeds-based firm John Fowler & Co. in the late 1930s. The locomotive was one of a number produced by the company and employed by the LMS and later BR's Engineering Department. For the majority of its career E.D. 1 was employed at Beeston Sleeper Works, but had a spell at West India Dock, London, and Castleton Engineer's Yard. The locomotive left service in 1962 and scrapped at Derby; it is seen at the works in mid-1955.

Derby 47166

LMS Fowler 0-6-0T locomotive no. 47166 was designed at Horwich Works with the guidance of T. F. Coleman, but the engine was erected at Derby in December 1928. Ten were completed to order no. 7137 at the works for working docks and other places that would benefit from their short 9ft. 6in. wheelbase. Between 1948 and withdrawal in May 1963, the engine was based at Birkenhead, Bidston, Sutton Oak, Bidston and Edge Hill. No. 47166 had been come from Bidston to be overhauled at Derby and is seen at the works on 13 May 1956.

Derby NCC 54

The Northern Counties Committee had been formed after the MR acquired the Belfast & Northern Counties Railway in 1903 and subsequently a number of engines were built for the railway at Derby. Shortly after Nationalisation two orders, no. 3283 and no. 4332, were placed for eight NCC H. G. Ivatt WT Class 2-6-4T locomotives, which were based on the LMS 2-6-4T engines, and these would prove to be the final locomotives constructed at Derby for the NCC. No. 54 was erected during 1950 as the first engine of the latter batch and is seen ready to be shipped to Northern Ireland. No. 54 was in service until April 1966.

CHAPTER FOUR

Drivers

Driver Tom Whit
Posing alongside Ivatt Class 4MT 2-6-0 no. 43040 is driver Tom Whit. The locomotive was constructed at Horwich Works in July 1949 and was one of fifty manufactured at the works to be fitted with a double chimney. However, due to problems caused by this feature it was removed and a new chimney arrangement was fitted to later locomotives and the earlier ones were also subsequently altered; no. 43040 was changed in May 1953 at Horwich. The locomotive is seen with driver Whit during October 1955 when it was working a local ballast train and had Bill acting as fireman. No. 43040 was withdrawn in November 1966 from North Blyth depot.

Driver C. Howlett
Ex-Annesley driver Charlie Howlett is seen in charge of a diesel locomotive.

Driver C. Willetts
Inside the cab of a Class 45 diesel-electric locomotive is driver Cliff Willetts.

Driver J. Fulture
At the controls of a Class 47 diesel-electric locomotive is driver Jess Fulture.

Driver G. Hayes
Driver George Hayes is seen working a Class 45 locomotive.

Driver J. Wilkinson
In charge of Class 45 no. 45126 at Nottingham holding sidings is driver Jim Wilkinson. The locomotive had been constructed at Derby Works in June 1961 as D32 and was initially based at Derby before working from Neville Hill and Holbeck sheds in Leeds. In the late 1960s and mid-1970s the locomotive was allocated to the Nottingham Division and Toton respectively. No. 45126 left service in April 1987 and was later scrapped.

Driver T. Carless
Driver Terry Carless poses for the camera inside the cab of a Class 45 locomotive.

Driver T. Whitmore
At the helm of this locomotive is driver Tom Whitmore.

Groups

NUR Meeting
Attendees of a National Union of Railwaymen, Nottingham Midland Branch, meeting pose for the camera.

Nottingham Railway Football Team
Seen in Melbourne Park, Nottingham, is the Nottingham Railway football team.

Railway Club Party
Gathered for a Railway Club party are: (on the left) E. Jackson, Ted Marshall, Ted Jones, (on the right) George Stone, Jack Green.

Mansfield

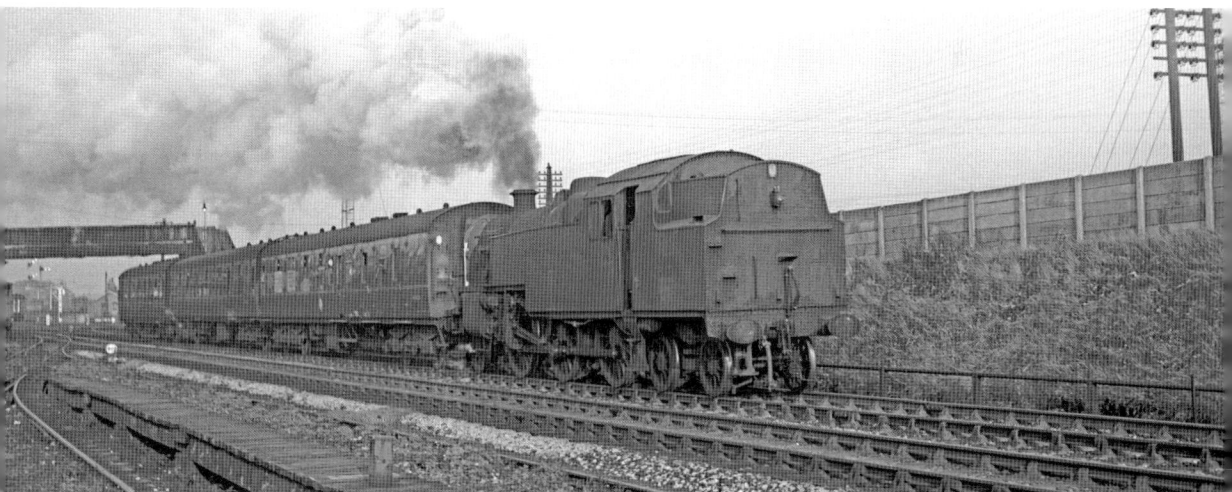

Mansfield 42588
Stanier 2-6-4T locomotive no. 42588 was constructed by the NBLC in October 1936. The engine's final allocation was to Nottingham and this lasted from September 1961 until withdrawn towards the end of October 1964. No. 42588 is photographed on the Nottingham – Mansfield line between Mansfield North (visible to the extreme left of the view) and Mansfield South signal boxes on 10 October 1964.

Mansfield South Junction 48317

When the 8F Class first appeared in the mid-1930s they were fitted with vertical throatplate boilers and had their smokebox-mounted tube cleaner cock in a low position near the smokebox saddle. When the engines with sloping throatplate boilers were introduced the cleaner cock was moved to a position above the boiler handrail, but from September 1957 a BR standard-type cleaner cock, which reverted to the original position, was fitted and the feed pipe was made external; no. 48317 has received the new arrangement. The locomotive was working from Kirkby-in-Ashfield when pictured on 10 October 1964 and would leave service in March 1968.

New Basford

New Basford 43152

Running northwards, tender first, on the former GCR line at New Basford is Ivatt 4MT 2-6-0 no. 43152. The locomotive was a Doncaster-built member of the class and had entered traffic in November 1951. It was fitted with tablet catchers on both sides of the tender tank for working on single line sections of the former Midland & Great Northern Join Railway line and a number of other engines were so fitted for working this line and for sections of Scottish Region track. The tender is also Doncaster-built as it has flush rivets in the centre for application of the emblem. No. 43152 was withdrawn from Colwick in January 1964; the engine is seen here on 14 September 1963.

New Basford 43145
Also erected at Doncaster, in September 1951, was Ivatt 4MT no. 43145, which is pictured from Perry Road bridge, New Basford, travelling southwards on 9 May 1964. Like no. 43145, the locomotive is fitted with tablet catchers to work the former M&GN line and was based at a number of sheds in that area. South Lynn, Yarmouth Beach, Melton Constable and Norwich accommodated the engine between 1951 and 1960, before no. 43145 arrived at Staveley in March. The locomotive later went to its final allocation at Colwick in September 1962 and was taken out of service during January 1965. The view is almost unrecognisable now as the cutting has been filled and houses sit on part of the land.

New Basford 45581

The NBLC's Hyde Park Works were responsible for the erection of no. 45581 and it was completed in September 1934 as the last of a batch of twenty-five engines built there. The locomotive had to wait until 30 March 1938 to be given its name – *Bihar and Orissa* – and was the last of the class to be a recipient of one. The plates were cast and fitted at St Rollox Works, but it had originally been the intention to have all the nameplates cast at Crewe. The plates from the former works differed from the standard by being lighter and having taller and slimmer lettering.

New Basford 45581

No. 45581 *Bihar and Orissa* has been caught twice by Bill on 15 August 1964 and this picture shows the engine after it has collected a passenger train from Nottingham that is possibly bound for Leeds. The locomotive had a long spell working from Farnley Junction shed, Leeds, and this lasted from September 1952 until no. 45581 left service in August 1966.

New Basford 46156

Fowler Royal Scot 4-6-0 no. 46156 *The South Wales Borderer* was built at Derby in October 1930 and was subsequently rebuilt with a taper boiler and new cylinders in May 1954. No. 46156 was attached to tender no. 9049 in March 1959, which had been involved in the tender exchanges between new Jubilees and Royal Scots in the mid-1930s. *The South Wales Borderer* had five months left in service when it was photographed at New Basford on 9 May 1964.

New Basford 60831

Built at Darlington in May 1938, Gresley V2 no. 60831 was loaned briefly to Heaton before spending eleven years working from Doncaster shed. The engine reallocated to Woodford in June 1949 and worked for just over eight years on the former GCR line from Woodford and Leicester sheds over two spells at each. No. 60831 appears to have a slight steam leakage from the monobloc cylinder casting, but it must have developed into a more serious problem from when pictured on 6 August 1955 to when the casting was replaced in mid-1957. The engine spent seven years at York before withdrawal on 6 December 1966 and was the last V2 to be taken out of service.

New Basford 45562
Heading an express passenger service through New Basford on 9 May 1964 is Stanier Jubilee no. 45562 *Alberta*. The locomotive spent a large portion of its career at Leeds Holbeck, but had recently transferred to Patricroft. *Alberta* has the distinction of being the last Jubilee to be withdrawn from service, after returning to Leeds Holbeck, on 4 November 1967.

New Basford 92132
BR Standard Class 9F 2-10-0 no. 92132 is noted in *British Railways Standard Steam Locomotives Volume Four* as heading the 10.48 a.m. passenger service from Eastbourne to Sheffield Victoria station, via the former GCR line, on the 15 August 1964 – the date it has been photographed at New Basford. The 9Fs were no strangers to passenger trains and from the late 1950s were used extensively during the summer holiday traffic. This arrangement lasted, in the most part, until the early 1960s, but 9Fs could still be seen on the GCR line with passenger trains up to the line's closure. No. 92312 was allocated to Annesley at the time of the picture and would be withdrawn from Carlisle Kingmoor in October 1967.

New Basford 61248

Thompson B1 4-6-0 no. 61248 *Geoffrey Gibbs* is also passing through New Basford on 15 August 1964. The locomotive had been allocated to Immingham since November 1959 and would move to Colwick in January 1965. The engine was condemned at the shed during November 1965.

New Basford 61264

B1 no. 61264 was erected at the NBLC's Queens Park Works in December 1947 and was initially allocated to Stratford before a quick transfer to Parkeston occurred. The locomotive was based at the shed for twelve years until moving to Colwick in November 1960. No. 61264 left service in November 1965, but saw further use as a departmental locomotive at Colwick until July 1967. Nine years later the engine was bought for preservation and no. 61264 returned to steam in 1997. Recently the locomotive has undergone a major refurbishment and can now be seen working on the North Yorkshire Moors Railway.

New Basford 61272

No. 61272 is seen moving empty coaching stock through New Basford on 6 August 1955 – the former LNER carriage sheds at New Basford can be seen on the left-hand side of the picture. The engine was based at Colwick at the time of the picture and had been there since February having moved from Leicester. Transfers to Annesley, Sheffield, King's Cross and Peterborough New England would occur before the locomotive was withdrawn in January 1965. The engine was another B1 pressed into Departmental service and this was its role until sold for scrap in January 1966.

New Basford 61281

No. 61281 was built at the NBLC's Queens Park Works in January 1948 and was allocated to Colwick shed between June 1958 and February 1966. No. 61281 is pictured at New Basford on 9 May 1964 and is displaying slight scale formation around the dome. The locomotive had last undergone a general repair in March 1962 at Doncaster and its final visit to works was from late August to early September 1963 at Darlington. The engine had received a fresh boiler at the general repair, no. 28859, which had previously been in the possession of no. 61168, and would be sold for scrap with no. 61268 in May 1966 to Birds, Long Marston.

New Basford 61768

Gresley K2 2-6-0 no. 61768 was constructed at the NBLC's Queens Park Works during August 1918 and would have begun service with a dark grey livery with white lining. The standard for the class later became black with red lining under the LNER, but a number of locomotives before Nationalisation, and indeed after, acquired a green livery before it was decided the class should have BR's mixed traffic livery applied. No. 61768 is seen in the latter livery running light engine at New Basford on 6 August 1955.

New Basford 64238

Gresley J6 no. 64238 was erected at Doncaster Works during October 1914 and was in service for forty-five years before being condemned in October 1959. During its career the locomotive was one of several class members to be fitted with an experimental top feed arrangement, which was covered by the dome being stretched forward on the boiler, and the apparatus was fitted to the locomotive from June 1917 until December 1926.

New Basford 70004

Running light engine through the cutting at New Basford towards Nottingham on 9 May 1964 is BR Standard Class 7 'Britannia' Pacific locomotive no. 70004 *William Shakespeare*. The locomotive emerged from Crewe Works in March 1951 and at first operated from Stratford before moving south of the River Thames to Stewarts Lane depot. No. 70004 had three spells at Willesden shed between December 1960 and January 1965 and was almost a year into its third allocation at this time. From Willesden the class were used on scheduled services and as replacements for diesel failures on the former GCR line and could also be employed on freight and goods services before they were transferred away at the start on 1965. *William Shakespeare* finished its career at Carlisle Kingmoor and was removed from service at the end of December 1967.

New Basford 90545

WD Austerity 2-8-0 no. 90545 is seen at New Basford on 14 September 1963. The locomotive appears to have come into some difficulty in tackling the rising gradient of 1 in 130 between Sherwood Rise tunnel, New Basford and Bagthorpe Junction and one of the mineral wagons may have been stopped by the catch points, where there is a gathering of people. The engine entered traffic from the Vulcan Foundry in July 1943 and was in service until October 1965. For the last five years of its life, the engine worked from Colwick.

New Basford 90629

No. 90629 was also erected by the Vulcan Foundry, but was produced slightly later in March 1944. The locomotive was allocated to York shed from October 1947 until August 1949 when it was reallocated to Colwick after undergoing a Heavy General Repair at Darlington Works. No. 90629 was then based at the shed until condemned in September 1965. The locomotive is on the former GCR line just to the north of New Basford station, which is just visible behind the signal on the right-hand side. The station was opened on 15 March 1899 and closed on 7 September 1964.

Nottingham

Nottingham London Road Junction Signal Box

A view inside Nottingham London Road Junction signal box with Les Mabbert at the levers. The junction was on the former MR line to Lincoln and provided a route in and out of Nottingham Midland station for trains coming off the line to Melton Mowbray and Kettering, which was opened in February 1880. The latter line was open until the late 1960s.

Nottingham Victoria Station 46101

The driver of this passenger train, which has stopped at Nottingham Victoria station, takes the opportunity to check the engine before the rest of the journey is undertaken. Fowler Royal Scot 4-6-0 no. 46101 *Royal Scots Grey* was constructed by the NBLC's Queens Park Works in September 1927 and was named the following April. *Royal Scots Grey* had reached Camden shed by the end of 1927 and this would remain the allocation until after Nationalisation when the locomotive would undergo twenty-four transfers to ten depots. No. 46101 ceased to be in service at the end of August 1963 after eight months at Annesley and was later scrapped by Slag Reduction Co., Rotherham.

Nottingham Victoria Station 46169

The GNR accepted the GCR's offer to share a station at Nottingham during March 1895 and Nottingham Joint Station Committee was formed in 1897 to oversee its construction. Edward Baker and W. H. Elwell, estate agents for the GCR and GNR respectively, were in charge of the purchase of the thirteen acres of land the station was to stand on, as well as the approach lines, and the process took three years and cost £473,000 to complete. Approximately 1,300 buildings were demolished and 600,000 cubic yards of earth displaced to allow Henry Lovatt to begin construction of the station, which had been contracted to cost £146,918. The station was opened to passengers on 24 May 1900 and initially the GCR and GNR could not agree on the name referring to it as Nottingham Central and Nottingham Joint respectively. Nottingham Victoria was agreed in June 1900 and was in use until closure to passengers on 4 September 1967 and then freight on 25 May 1968. No. 46169 *The Boy Scout* is seen at the station on 2 March 1963.

Nottingham Victoria Station 64354

Former GCR 9J Class 0-6-0 locomotive no. 177 became LNER no. 5177, quite some time after Grouping, in November 1926 and in the 1946 scheme became LNER no. 4354 in June, then BR no. 64354 was carried from August 1948. The engine would have originally been paired with a 3,250 gallon tender with a water scoop. However, in this photograph of the locomotive at Nottingham Victoria station on 29 May 1957, a Robinson 4,000 gallon tender is in use. Only a couple of weeks had elapsed since the locomotive was transferred to Colwick from Annesley, but it would only be a relatively brief stay at the former, as Tuxford would take charge of the engine from February 1958. Langwith Junction, Staveley and Retford would be visited before September 1962 saw no. 64354 leave service as the last member of the class.

Nottingham Victoria Station 46111

Nottingham Victoria station was designed by local architect Albert Edward Lambert in the Renaissance style and utilised bricks and Darley Dale stone. There were two island platforms (1,250 and 1270 yards long) with two bay platforms at each end and they were covered by a three-canopy glass roof, which was supported by iron columns 42ft. 6in. tall. Glass would have originally been present in the frontage, but it has since been replaced by corrugated sheeting. No. 46111 *Royal Fusilier* (seen here minus its nameplate) was erected in October 1927 and relieved of its duties in September 1963.

Nottingham Victoria Station 61181

No. 61181 was built by the Vulcan Foundry in July 1947 and after being delivered to Gorton was transferred to Sheffield and spent the best part of its career working from the depot. By 1948 there were ninety-three Thompson B1s working from former GCR sheds, with Sheffield Darnall (fifteen) and Gorton (twenty) having the largest number at hand. The contingent available for work at Sheffield rose steadily through the 1950s and by 1961, forty-four engines were at the shed to work a wide range of services including; the expresses to London (generally only as far as Leicester), cross-country services and both long and short distance goods trains.

Nottingham Victoria Station 61889

No. 61889 was paired with a Group Standard tender with straight sides and a coal and water capacity of 7 tons 10 cwt and 4,200 gallons respectively when new at Doncaster in December 1929. The type is still present with the locomotive, which is seen at Nottingham Victoria station during May 1961. The tender also appears to have undergone a modification that was applied to the Group Standard tenders during the 1950s. This involved the fitting of a new division plate that was attached to the top of the tank 1ft. 10½in. further forward and was slightly more than 11in. higher higher than the previous arrangement. The engine was Colwick-based at this time, but would be withdrawn from Immingham in November 1962.

Nottingham Victoria Station 62203

The GNR Ivatt Class D1 4-4-0s constructed after 1909 (LNER D1) were different from the earlier engines as they were equipped with Schmidt superheaters and 8in. piston valves. Only fifteen locomotives of the design were constructed – all at Doncaster – and no. 62203 was the second to be completed, as GNR no. 52, in April 1911. Of the seven D1s that were still in service at Nationalisation only three received their BR number and no. 62203 was one when it was applied in March 1948. The engine is photographed at Nottingham Victoria in late 1949/ early 1950, before it was withdrawn in August 1950.

Nottingham Victoria Station 2140

H. A. Ivatt produced the design for the GNR's first 4-4-0 in the mid-1890s and the first engine, GNR no. 400, entered traffic in December 1896. Subsequently a further fifty were built at Doncaster Works and no. 1348, LNER no. 2140, was completed in December 1898; the locomotives were classified D2 by the GNR. However, Gresley fitted a larger boiler with a shorter firebox to the class from 1912, bringing about a reclassification of the modified engines to D3, which was also the classification applied to the class after Grouping; no. 1348 received the new boiler in May 1915. No. 2140 was sixteen months away from withdrawal in this picture, which was taken on 6 February 1949.

Nottingham Victoria Station 70045

BR Standard Class 'Britannia' Pacific locomotive no. 70045 *Lord Rowallan* was built at Crewe Works during June 1954 at a cost to BR of £23,446, which was just over a £3,000 increase in the construction cost of no. 70000 *Britannia* in 1951. Nos. 70043 – 70054 were nameless when entering traffic and all but one engine were subsequently bestowed with one. Lord Rowallan, who was Chief Scout of The Boy Scouts Association (later The Scouts Association) from 1945-1959, named this engine at a ceremony at Euston station on 16 July 1957. No. 70045 was a short time away from moving to Neasden in this image of the locomotive taken at Nottingham Victoria during August 1961.

Nottingham Victoria Station 46126

Another Fowler Royal Scot is at the head of a passenger train at Nottingham Victoria station and this image dates from 5 August 1963. No. 46126 *Royal Army Service Corps* was built at the NBLC's Hyde Park Works in September 1927 at a cost of £7,744, which was approximately the cost of each of the first fifty engines ordered from the company in 1926 as part of the LMSR's 1927 locomotive building programme. No. 46126 was named *Sanspareil* in the early part of 1928 following the policy of naming some of the class after early locomotives. The engine became *Royal Army Service Corps* in June 1936 after a change was implemented earlier in the decade to regiments in the army. No. 46126 was, like its classmate *Royal Scots Grey*, working from Annesley shed and would be condemned while there in October 1963.

Rigley's Wagon Works

Rigley's Wagon Works 63925

Due to the success of the first O2 Class 2-8-0 engine, GNR no. 461, which used Gresley's conjugated valve gear, Gresley decided to produce more three-cylinder locomotives and a further ten O2 Class engines were ordered from Doncaster Works. However, this order was later transferred to the NBLC's Atlas Works and no. 63925 was the fourth engine of the batch to be completed in May 1921. The batch featured a few differences from the pioneer locomotive and these included; alterations to the conjugated valve gear, steam pipes redesigned and increased cylinder diameter to 18½ inches. No. 63925 was withdrawn in September 1963 and is seen on 7 November 1964 at William Rigley & Sons Ltd Wagon Works, Bulwell Forest, awaiting scrapping.

Rigley's Wagon Works 63939

Gresley O2 Class 2-8-0 no. 63939 is pictured slightly earlier at W. Rigley's Wagon Works on 28 August 1964. The locomotive had also been withdrawn in September 1963, which was a particularly sombre month for the class as no fewer than thirty engines were condemned. Of this number twenty-one were sent to Rigley's Wagon Works to be scrapped as well as another class member that had been withdrawn in May 1963.

Rigley's Wagon Works 63985

No. 63985 was one of the final three class members that were erected at Doncaster Works in January 1943. The O2 Class had been designated a Standard Design in the late 1920s and a total of thirty-six were built to the specifications between 1932 and 1943. Some of the changes from the GNR O2s that were implemented included; left-hand drive, Group Standard tenders, long travel valve gear, side window cabs (that were also larger), lower dome cover and altered sanding and lubricating arrangements. In addition, the O2s produced in the 1940s had an improved superheater arrangement, larger coupled wheel axle journals and vacuum brakes. No. 63985 was also one of a number of O2s that were fitted with AWS equipment and the battery box is visible below the cab. The engine left service in September 1963 after a brief allocation to Retford.

CHAPTER FIVE

Ruddington

Ruddington 44821

At Ruddington on the former GCR line, which ran parallel to the village located to the south of Nottingham, during June 1962 is Stanier Class 5 no. 44821. The locomotive was built at Derby Works in December 1944 and was one of a number in the batch (LMS nos. 4807-4825) that were fitted with bronze axlebox guide liners. They replaced the white metal type used previously in an unsuccessful attempt to reduce wear. The bronze axlebox liners would later be replaced by manganese steel liners after they became standard towards the end of the decade. No. 44821 was withdrawn in June 1967 from Crewe South shed.

Ruddington 61136

Ruddington station, formerly on the west side of the village, was opened by the GCR on 15 March 1899 with a small island platform and was in use until 4 March 1963. Thompson B1 no. 61136 is photographed in charge of a passenger service nine months before the station's closure. The engine was constructed at the NBLC's Queens Park Works during March 1947 and during the 1950s was based at Neasden, Leicester and Woodford. The latter had been reached a few months before the picture was taken and would see the locomotive leave service in November 1962.

Ruddington 64749

Gresley J39 0-6-0 locomotive no. 64749 was erected at Darlington Works, which built 261 of the 289 engines of the class, in August 1928. The engine was originally paired with a 3,500 gallon Group Standard tender that had a coal capacity of 5½ tons and is still paired with the type when photographed on 20 June 1961. No. 64749 is a distance from home ground here as it was an Ardsley engine at this time and had been since May 1948. Withdrawal from the shed occurred in November 1962. Note the lumps of coal on the roof of the track-side hut on the right, which have perhaps been 'donated' by passing engines.

Ruddington 61455

NER Raven S3 Class, LNER B16 Class 4-6-0 no. 61455 was erected at Darlington in October 1923. The NER ordered the final thirty-five engines of the class from the Works in January (ten), March (ten) and December 1922 (fifteen), but only three were completed before Grouping. The locomotive was rebuilt to class part two specifications in November 1939 and in July 1944 a diagram 49A boiler was fitted. The engine remained in this state until it left traffic during September 1963. *Locomotives of the LNER Part 2B* (1975) notes that no. 61455 accumulated the highest mileage of the class with 1,159,799 miles.

Ruddington 70037

BR 'Britannia' Pacific no. 70037 *Hereward the Wake* is seen at Ruddington on 5 June 1962 with a freight train. The engine was completed at Crewe Works in December 1952 and has undergone two modifications since then. One is the removal of the handrails on the smoke deflectors and their replacement by two hand holes on each side. AWS has also been fitted to the engine and this feature was present on the locomotive from October 1960. Working from Immingham at the time of the picture *Hereward the Wake* was withdrawn from Carlisle Kingmoor in November 1966 after nine months of being stored as unserviceable.

Ruddington 73157

This image was captured from Station Road bridge, now Clifton Road bridge, on 21 June 1961 looking north along the former GCR route and shows BR Standard Class Five no. 73157 heading south with a London Marylebone express. Built at Doncaster Works in December 1956 the locomotive was one of five allocated to Neasden when new (nos. 73155-73159). The engine was transferred away twice before returning to the shed for a final two-year spell from June 1960. No. 73157 was at Patricroft when condemned in May 1968.

Ruddington 92012

From the outset of BR's standardised locomotive programme the three men at the forefront – R. A. Riddles, R. C. Bond and E. S. Cox – debated whether the standard freight engine should have a 2-10-0 or 2-8-2 wheel arrangement. The proponents of the 2-8-2 were R. C. Bond and E. S. Cox, who thought the higher boiler capacity would be ideal for BR's proposed fast freight service. However, the use of continuous brakes on all wagon stock, which would have been required for it, was not a feasible option at the time and nullified the argument for the 2-8-2. No. 92012 is seen with a through freight at Ruddington on 5 June 1962.

Ruddington 92032

The 2-10-0 wheel arrangement was adopted despite a large amount of design work being carried out on the 2-8-2 engine. R. A. Riddles had argued that the 2-10-0 locomotive had a better tractive effort and adhesion factor than the 2-8-2 and would improve timings by utilising these strengths when starting and on adverse gradients rather than running at high speed. No. 92032 was erected at Crewe in November 1954 and left service in April 1967.

Ruddington 92091

Between January 1954 and March 1960 251 9F Class locomotives were constructed at Crewe and Swindon. No. 92091 was completed at the latter in November 1956 and after initially working from Doncaster, moved to Annesley in March 1957 and spent eight years working from the depot. Withdrawal from Carnforth occurred in May 1968.

Ruddington 92093

Between February and March 1957 eighteen BR 9F 2-10-0s (nos. 92067-92076 and nos. 92087-92094) were moved from Doncaster to Annesley as part of a restructuring of the Eastern Region's freight locomotives. Annesley already had four 9Fs at this time and also acquired six from New England in addition to two new engines from Swindon. The 9Fs were used on the Annesley to Woodford Halse freight service replacing the O1 Class 2-8-0s. No. 92093 was condemned at Carlisle Kingmoor in September 1967.

Sheet Stores Junction

Sheet Stores Junction 44804

Sheet Stores Junction was one of the many connections in the area around Trent station, where the lines from Nottingham, Derby, Leicester and Chesterfield converged. The station opened in the early 1860s after a major alteration of the connections between the lines; Sheet Stores Junction was opened to provide a link from Trent Station South Junction to the section of track between the Derby line and the Leicester line. Between the late 1860s and early 1870s a line was built from Sheet Stores Junction to the Birmingham & Derby Junction Railway line at Stenson Junction, which allowed trains, mainly freight, to bypass Derby. No. 44804 is pictured at Sheet Stores Junction on 6 June 1962.

Spondon

Spondon, Albert Looms Ltd Nos. 50818 and 50831

Two former Lancashire & Yorkshire Railway 2-4-2T locomotives have arrived at Albert Looms Ltd, Spondon, Derby, to be broken up. They were members of John Aspinall's K2 Class (or Class 5), which were constructed between 1889 and 1911 and totalled 309 engines. No. 50818, L&YR no. 730, was erected at Horwich Works in September 1898, while no. 50831, L&YR no. 316, was constructed at the works in November of the same year: the latter locomotive was later modified to have a larger tank and bunker. At Grouping 278 engines passed to the LMS (the reduction in total being due to rebuilding, with the first not withdrawn until 1927) then being classified 2P and 110 remained to enter service for BR. No. 50818 left traffic in October 1958 from Sowerby Bridge shed, while no. 50831 followed in the following month from Low Moor depot.

Toton

Toton 41966

LT&SR Whitelegg 79 Class 4-4-2T locomotive no. 80 *Thundersley* was built by Robert Stephenson & Co. Ltd during May 1909. In 1912 the LT&SR was acquired by the MR and the locomotive was taken into their stock, becoming no. 2177. The engine later received LMS no. 2148 in 1930 and BR no. 41966 after Nationalisation, which appears to have been painted on rather than applied in transfers. No. 41966 arrived at Toton during February 1953 and the allocation lasted until withdrawal in June 1956. Subsequently the locomotive was preserved and is currently on loan to Bressingham from the National Railway Museum. Toton shed's water tank is seen in the background, which was on the north-west side of the site, and originally drew its supply from Derby Canal, but from 1956 the water came from the Erewash Canal.

Toton 41947

This LMS 4-4-2T engine was erected at Derby in June 1927 as part of order no. 6751 for ten locomotives that were to be used on the former LT&SR line. Derby had produced ten previously in 1923 and would construct a final batch of ten in 1930. The locomotive was allocated to Toton between February 1956 and November 1960 when it was the last of the class to be withdrawn. No. 41947 is pictured at Toton during August 1958.

Toton 47004

Former LMS 0F 0-4-0ST locomotive no. 47004 is some distance from its home shed at Hasland, near Chesterfield, when pictured at Toton during July 1958. The engine was based at the former depot throughout its time working for BR and was relieved of its duties while there in January 1964. No. 47004 has the latter BR emblem applied and it is facing the incorrect way.

Toton 41966

No. 41966 is seen from the front end at Toton shed in the mid-1950s. The BR number has also been applied to the smokebox by hand, rather that number plate, in addition to Toton's 18A shed code. Toton was 18A from 1935 until September 1963 when it became 16A.

Toton 47247

MR Johnson 0-6-0T locomotive no. 47247 was built by the Vulcan Foundry in June 1902. Construction of the class at the works had begun in 1898 and a total of sixty had been completed by 1902. Thirty of these were fitted with condensing apparatus to work in the London area and no. 47247 still has it fitted, despite working outside the capital since at least Nationalisation. The engine had two spells at Toton and is pictured in 1955 during the second, which lasted from December 1954 until the locomotive left service in August 1959.

Toton 47972
LMS Garratt 2-6-0+0-6-2 locomotive no. 47972 was constructed by Beyer, Peacock & Co. in September 1930. Toton was the main residence for the class and the main task for them was transporting coal from the adjacent marshalling yard to Brent Sidings, London. No. 2 and no. 3 sheds at the depot were modified to house the locomotives by having a road running the length of the two buildings installed to accommodate them. No. 49792 is seen at Toton on 6 October 1955, but the engine was later moved to Hasland and was condemned there in April 1957.

Toton
An unidentified LMS Garratt 2-6-0+0-6-2 locomotive is seen leaving Toton marshalling yard during the mid-1950s. All of the thirty-three class members were constructed at Beyer, Peacock & Co., with three entering traffic initially in 1927 and thirty following in 1930. Thirty-two of the Garratts were fitted initially with a coal bunker, but in the early 1930s thirty were fitted with a rotating coal bunker in addition to one carrying the equipment when new. Of these, two built in 1927 remained without the rotating bunker during their time in traffic – nos 47998 and 47999 – and it can be assumed that one of them is the locomotive in this picture.

Toton 48088

No. 48088 was one of 51 8Fs requisitioned by the WD between November and December 1941 for use in the Middle East. However, while in transit, a storm in the Irish Sea damaged the locomotive and a number of others (four strapped to the deck were lost at sea) forcing the ship back to port at Glasgow. The engine was repaired at St Rollox, but was not re-dispatched and was taken on loan by the LMS in May 1942 and then bought back a year later. No. 48088 was allocated to Mansfield when seen at Toton and would be withdrawn from Buxton in December 1966.

Toton 58171

MR Johnson 1142 Class 0-6-0 locomotive no. 58171 was erected by Neilson & Co. in June 1876 as MR no. 1230. The engine was fitted with a G6 type boiler in February 1920, but the original type B boiler had been working at 160 psi since February 1916. No. 58171 worked at Toton from at least 1950, having been at Leeds Holbeck for Nationalisation. The locomotive left service from Chester in August 1959.

Toton 92058

BR Standard Class 9F 2-10-0 no. 92058 stands on Toton shed's yard awaiting its next duty during 1957, while an unidentified 9F fitted with a Crosti boiler is also seen behind the engine. No. 92058 was one of ten constructed between August and October 1955 at Crewe Works for the London Midland Region and at the start of their time in traffic these locomotives were based at Toton. The 9Fs at the shed generally replaced the Garratt locomotives on the coal workings to and from the London area. No. 92058 was removed from traffic in November 1967.

Toton 92102

Another 9F pictured at Toton during 1957 is no. 92102. The locomotive was also allocated to Toton from new, but in March 1958 a move to Leicester Midland depot transpired and in April 1965 a final reallocation to Birkenhead occurred. Toton shed had forty-seven of the 9F Class allocated for varying periods, with the longest allocation belonging to no. 92078, which arrived new in March 1956 and left in March 1965. No. 92102 left service in November 1967; Toton had closed to steam a year earlier.

Toton 48772

An LNER-built example of Stanier's 8F Class poses for this portrait at Toton on 10 May 1955. No. 48772, LNER no. 3167 and later 3567, was built at Doncaster Works in June 1946 narrowly avoiding being the 2000th locomotive to be constructed at Doncaster, carrying works no. 2001. The locomotive was taken into LMS stock in September 1947 and was based at Crewe South, Widnes and Wellingborough before arriving at Toton in December 1953. The 8Fs at the shed were also used on freight and mineral trains from Toton marshalling yard. No. 48772 was condemned at Staveley in January 1964.

Trowell Summit

Trowell Summit 44853

Tackling the gradients to Trowell Summit on the Radford to Trowell Junction line is a resplendent Stanier Class Five locomotive, no. 44853, which was constructed at Crewe Works in November 1944. The line linked the Mansfield line with the Erewash Valley line and was built by the MR in the early 1870s, opening on 1 June 1875. No. 44853 was a long-term Leeds Holbeck resident and was withdrawn from there in June 1967. The train is passing under Coventry Lane bridge.

Trowell Summit 45529
The Radford to Trowell Junction line was authorised by the Midland Railway Branches Act of June 1866. The route was 4 miles 77 chains long and Eckesley & Bayliss tendered a price of approx. £35,064 for the line's construction, which began in mid-1870. Fowler Patriot no. 45529 *Stephenson* is pictured here during July 1956 in its rebuilt form; this taking place in July 1947. The locomotive was built at Crewe Works in April 1933 and had originally been named *Sir Herbert Walker K.C.B.*, but this was only carried until 1937 and *Stephenson* was carried from July 1948. The engine ceased to be in service during February 1964 at Annesley.

Watnall

Watnall 64747
Running light engine on the former GNR Derbyshire & Staffordshire Extension line between Basford and Kimberley at Watnall on 24 September 1957 is Gresley J39 Class 0-6-0 no. 64747, which was erected at Darlington Works in August 1928. Behind the locomotive is Nuthall Sidings and the branch line to Watnall colliery, which was sunk in 1875 and closed in 1952. No. 64747 spent twenty-six years working in the Nottingham area at Annesley and Colwick and was coming to the end of its allocation to the latter at this time. The locomotive was withdrawn from traffic in November 1962 while at Woodford, but saw further use at the shed as a stationary boiler.

Watnall 90288

WD Austerity 2-8-0 no. 90288 hauls a load of coal on the line at Watnall on 6 June 1957. The engine was erected by the NBLC in February 1943 as WD no. 7415. No. 90288 had undergone a Heavy General repair over a year prior to this picture being taken and another HG repair would be just under two years away. Gorton was the works responsible in both instances and it had maintained the locomotive up to withdrawal in September 1962 from Frodingham depot.

Watnall 61768

No. 61768's allocation to Colwick lasted from August 1946 until condemned in January 1959 after travelling to Darlington for repair. The engine was somewhat a nomad between 1935 and 1940 as nine moves were undertaken taking it to; Norwich (twice), Yarmouth, Ardsley (three times), Bradford (twice) and Peterborough New England. Doncaster was allocated the locomotive for the longest time, nineteen years, but this was split over three periods of seven, nine and three years between 1918 and 1945. The engine had less than a year left in service when pictured at Watnall on 11 March 1958.

Watnall 63839

This view was taken on the east side of Fan Bridge (no. 41) over the Derbyshire & Staffordshire Extension line. The line running on the right-hand side of the picture is to the Stanton Ironworks Company's sand quarry, which was opened on the 'down' side of the line towards the end of the nineteenth century. In the 1930s the 'up' side of the line began to be worked and the bridge for the wagons from that side to the sidings can be seen in the distance. O4/7 Class 2-8-0 no. 63839 could be heading a train bound for Stanton Ironworks on 24 September 1957.

Watnall 64974

Moving west, on the same date, along the line to bridge no. 42, which was a footbridge for a footpath leading from Bulwell cemetery in Nuthall, and Gresley J39 Class no. 64974 is seen at the front of a coal train. When built in March 1941 at Darlington, the locomotive was paired with a second-hand tender, no. 8587, from withdrawn D17 Class 4-4-0 no. 1636 as a means of saving material during the war. No. 64974 was paired with two other 3,650-gallon NER tenders while in service; no. 8597 between May 1946 and February 1959 and no. 8924 to being condemned in August 1960.

Watnall 64195

Gresley J6 0-6-0 no. 64195 was constructed at Doncaster Works in May 1913 with; 5ft. 2in. diameter driving wheels; two inside cylinders measuring 19in. diameter by 26in. stroke; Stephenson motion with 8in. piston valves; 4ft. 8in. diameter boiler with an 18 element Robinson superheater. The only subsequent alteration to the locomotive's design was the fitting of a smaller chimney to bring the engine within the Loading Gauge during April 1941. No. 64194 left service from Colwick in January 1958.

Watnall 64755

Between the cessation of construction of the first batch of J39s in September 1927 and the beginning of the second batch some slight alterations were made to some features used on the engines. Screw reverse, Owen regulator, exhaust steam injectors and mechanical lubricator for the axleboxes and cylinders replaced steam reverse, slide regulators (installed on the majority of the first batch), live steam injectors and sight feed lubricator for the cylinders. No. 64755 was fitted with these modifications when entering traffic from Darlington in August 1928.

Watnall 67760

Working a local service on 24 September 1957 is Thompson L1 Class 2-6-4T no. 67760, which was erected at the NBLC's Queen Park Works in December 1948. The company produced thirty-five members of the class, while Robert Stephenson & Hawthorn contributed a further thirty-five and Darlington twenty-nine. Doncaster constructed the prototype, but no further engines to the design. The last five years of no. 67760's career was spent working from Colwick and the engine ceased to be in service from August 1961. In the background is footbridge no. 42.

Watnall 90080

Seen with a train of empty wagons at Watnall on 30 May 1965 is WD Austerity 2-8-0 no. 90080. The locomotive was built by the NBLC during March 1944 and the class were originally paired with 5,000-gallon tenders that had a coal capacity of nine tons. No. 90080 was paired with tender (BR) no. 81 from at least Nationalisation until August 1954 when the first switch occurred and tender no. 505 (from no. 90504) was substituted. Five months on from this picture being taken the engine would be given tender no. 657 and this would remain with the engine until it was condemned in February 1966.

Watnall 90189

Colwick-allocated Austerity no. 90189 is photographed at Watnall on 24 September 1957. The locomotive was twenty months into the residency having arrived from Mexborough, which had been a long-term home for the engine from October 1947 to the transfer in January 1956. No. 90189 was moved to Frodingham in February 1960 and then Barrow Hill in January 1964. The final month of the engine's career was spent at Langwith Junction and it was marked to be scrapped in November 1965.

Watnall 90669

Eight years on from the picture of O4/7 no. 63839 (p.119) taken from the east side of Fan Bridge and the line to Stanton Ironworks Company's quarry has now been removed and the land has become rather overgrown; the quarry was closed *c.* 1960. No. 90669 is hard at work moving coal around the area in mid-1965, but was a new arrival to Colwick from Frodingham. The locomotive would only be at work for a short while longer and was withdrawn in February 1966.

Watnall 92106

BR 9F Class locomotive no. 92106 entered traffic from Crewe Works in September 1956 and began its service life at Wellingborough shed. A number of tender varieties were paired with the 9F Class and the type typically depended on the region allocated. The BR1C tender, with a 4,725-gallon water capacity and space for nine tons of coal, was standard for the London Midland Region and no. 92106 received one when completed for work in that area. No. 1316 was the original coupling, but in May 1962 no. 1144 was acquired and before withdrawal in July 1967 no. 1328 was attached. No. 92106 is pictured moving empty wagons on 30 May 1965.

Wilford

Wilford 42556

LMS Stanier Class 4P 2-6-4T locomotive no. 42556 was constructed by the NBLC during July 1936. The company produced 73 of the locomotives in the class between 1936 and 1937, while the other 133 examples were completed at Derby Works from 1937 to 1943. No. 42556 is seen just to the south of Nottingham city centre at Wilford on the former GCR main line, towards the end of June 1962. The engine was well into its final allocation to Leicester and is likely to be working a local train between the two cities. No. 42556 was deemed surplus to requirements in July 1963.

Wilford 44847

At the head of an 'up' express passenger service on 12 August 1964 is Stanier 'Black' 5 no. 44847 and it is just passing Wilford Brick Works, which can be partially seen on the right-hand side of the picture. The engine was built at Crewe during November 1944 and was the second of seven to emerge from the Works in the month. The locomotive has the later tube cleaner cock arrangement in addition to the new top feed with 'top hat' cover and lower top lamp bracket. No. 44847 left service in November 1966.

Wilford 44941

Stanier Class Five 4-6-0 is in the twilight of its career in this picture taken on 28 May 1966 as it would be withdrawn in November from Colwick. The locomotive had been at the shed since January and has the abbreviation 'COLK' painted on the smokebox door instead of the shed plate. No. 44941 had entered traffic from Horwich Works in December 1945 and had narrowly missed out on the fitting of a few new features including a rocking grate and hopper ashpan, which was present on the following locomotive no. 44942.

Wilford 92072
An 'up' train of coal wagons passes Wilford Brick Works on 12 August 1964 with BR 9F 2-10-0 no. 92072 at the helm. The locomotive was built at Crewe Works in February 1956 for the Eastern Region and was initially allocated to Doncaster before being moved to Annesley. BR1F tender no. 1161 was present with the engine throughout its time in service and would have been fitted with Briquette Tube Feeder for water softening purposes as the water used at Annesley was very hard. No. 92072 moved to Kirkby-in-Ashfield in June 1965, but was placed in store at the shed until withdrawal in January 1966.

Williamthorpe Colliery

Williamthorpe Colliery 47289
Receiving some water and attention at Williamthorpe Colliery is Fowler 3F 0-6-0 no. 47289. The locomotive was one of the fifty 3F 0-6-0s authorised as part of the 1924 building programme and that were to be produced at the Vulcan Foundry (20), NBLC's Queens Park Works (15) and Hunslet Engine Co. (15). No. 47289 was erected by the NBLC in October 1924 and had a long career, being one of the last three to be withdrawn from service at the start of October 1967. The engine had been allocated to Westhouses shed at the start of the year and had worked from the depot to Williamthorpe Colliery to be pictured on 15 August 1967.

Williamthorpe Colliery 47383

Nos 47383 and 47629 were the other two members of the 3F Class to be withdrawn in October 1967 and are also seen at Williamthorpe Colliery in August 1967. No. 47383 was built by the Vulcan Foundry in October 1926 as LMS no. 16466 (no. 7383 from 23 July 1935), while no. 47629 was erected by William Beardmore & Co. in November 1928 as LMS no. 16712 (no. 7629 from 3 October 1934). One of the differences between the locomotives concerns the brakes and no. 47383 is fitted with steam and vacuum, whereas no. 47629 has steam only. After leaving service, no. 47383 was preserved and is currently based at the Severn Valley Railway.

Williamthorpe Colliery 68012

No. 68012 was the last of the LNER's J94 Class in service when photographed at Williamthorpe Colliery on 15 August 1967, but would soon be withdrawn in October. The engine carries the unofficial name 'Rasputin', which has been applied utilising the R.A.5 route availability classification for the first three letters. No. 68012 was fitted with an extended bunker in November 1947. However, this was removed in the late 1950s when the locomotive was sent for use on the Cromford and High Peak line as this would make the task of coaling easier at one of the line's depots.

Bibliography

Anderson, P. Howard. *Forgotten Railways: The East Midlands*. 1973.

Casserley, H. C. and S. W. Johnston. *Locomotives at the Grouping No. 3: London Midland and Scottish*. 1966.

Clough, David N. *British Rail Standard Diesels of the 1960s*. 2009.

Dow, George. *Great Central Volume One: The Progenitors, 1813-1863*. 1985.

Dow, George. *Great Central Volume Two: Dominion of Watkin, 1864-1899*. 1985.

Dow, George. *Great Central Volume Three: Fay Sets the Pace, 1900-1922*. 1985.

Griffiths, Roger and John Hooper. *Great Northern Railway Engine Sheds Volume 2: The Lincolnshire Loop, Nottinghamshire & Derby*. 1996.

Griffiths, Roger and Paul Smith. *The Directory of British Engine Sheds and Principal Locomotive Servicing Points: 2*. 2000.

Grindlay, Jim. *British Railways Steam Locomotive Allocations 1948-1968: Part Three London Midland and Scottish Regions 40001-58937*. 2008.

Hawkins, Mac. *The Great Central: Then and Now*. 2002.

Haresnape, Brian. *Fowler Locomotives: A Pictorial History*. 1997.

Haresnape, Brian. *Stanier Locomotives: A Pictorial History*. 1974.

Henshaw, Alfred. *The Great Northern Railway in the East Midlands*. 1999.

Henshaw, Alfred. *The Great Northern Railway in the East Midlands No. 3*. 2000.

Hooper, J. *The WD Austerity 2-8-0: The BR Record*. 2010.

Hunt, David, Fred James and Bob Essery with John Jennison and David Clarke. *LMS Locomotive Profiles No. 5: The Mixed Traffic Class 5s – Nos. 5000-5224*. 2003.

Hunt, David, Fred James and Bob Essery with John Jennison and David Clarke. *LMS Locomotive Profiles No. 6: The Mixed Traffic Class 5s – Nos. 5225-5499 and 4658-4999*. 2004.

Hunt, David, John Jennison, Fred James and Bob Essery. *LMS locomotive Profiles No. 8: The Class 8F 2-8-0s*. 2005.

Hunt, David, John Jennison, Bob Essery and Fred James. *No. 10: The Standard Class 4 Goods 0-6-0s*. 2007.

Hunt, David, Bob Essery and John Jennison. *LMS Locomotive Profiles No. 14: The Standard Class 3 Freight Tank Engines*. 2010.

Hurst, Geoffrey. *The Midland Railway Around Nottinghamshire Volume 1*.

Kingscott, Geoffrey. *Lost Railways of Nottinghamshire*. 2005.

Quick, Michael. *Railway Passenger Stations in Great Britain: A Chronology*. 2009.

Radford, Brian. *Rail Centres: Derby*. 2007.

Radford, J. B. *Derby Works and Midland Locomotives*. 1971.

RCTS. *Locomotives of the LNER: Part 2B Tender Engines – Classes B1 to B19*. 1975.

RCTS. *Locomotives of the LNER: Part 3B Tender Engines – Classes D1 to 12*. 1980.

RCTS. *Locomotives of the LNER: Part 4 Tender Engines – Classes D25 to E7*. 1968.

RCTS. *Locomotives of the LNER: Part 5 Tender Engines – Classes J1 to J37*. 1984.

RCTS. *Locomotives of the LNER: Part 6A Tender Engines – Classes J38 to K5*. 1982.

RCTS. *Locomotives of the LNER: Part 6B Tender Engines – Classes O1 to P2*. 1991.

RCTS. *Locomotives of the LNER: Part 6C Tender Engines – Classes Q1 to Y10*. 1984.

RCTS. *Locomotives of the LNER: Part 7 Tank Engines – Classes A5 to H2*. 1991.

RCTS. *Locomotives of the LNER: Part 8A Tank Engines – Classes J50 to J70*. 1970.

RCTS. *Locomotives of the LNER: Part 8B Tank Engines – Class J71 to J94*. 1971.

RCTS. *Locomotives of the LNER: Part 9A Tank Engines – Classes L1 to N19*. 1977.

RCTS. *Locomotives of the LNER: Part 9B Tank Engines – Classes Q1 to Z5*. 1977.

RCTS. *Locomotives of the LNER: Part 10A Departmental Stock, Locomotive Sheds, Boiler and Tender Numbering*. 1991.

RCTS. *The Railways of Nottingham*. 1969.

RCTS. *British Railways Standard Steam Locomotives Volume 1: Background to Standardisation and the Pacific Classes*. 1994.

RCTS. *British Railways Standard Steam Locomotives Volume 2: The 4-6-0 and 2-6-0 Classes*. 2003.

RCTS. *British Railways Standard Steam Locomotives Volume 4: The 9F 2-10-0 Class*. 2008.

Sixsmith, Ian. *The Book of the Ivatt 4MTs: LM Class 4 2-6-0s*. 2012.

Sixsmith, Ian. *The Book of the Royal Scots*. 2008.

Summerson, Stephen. *Midland Railway Locomotives Volume Four*. 2005.

Townsin, Ray. *The Jubilee 4-6-0's*. 2006.

Vanns, Michael A. *Rail Centres: Nottingham*. 2004.

Vanns, Michael A. *An Outline History of the Railways of Nottinghamshire*. 2010.

Walmsley, Tony. *Shed by Shed Part One: London Midland*. 2010.

Walmsley, Tony. *Shed by Shed Part Two: Eastern*. 2010.

Wrottesley, John. *The Great Northern Railway Volume 2: Expansion & Competition*. 1979.

Yeadon, W. B. *Yeadon's Register of LNER Locomotives Volume Four: Gresley V2 and V4 Classes*. 2001.

Yeadon, W. B. *Yeadon's Register of LNER Locomotives Volume Six: Thompson B1 Class*. 2001.

Yeadon, W. B. *Yeadon's Register of LNER Locomotives Volume Nine: Gresley 8-Coupled Engines Classes O1, O2, P1, P2 and U1*. 1995.

Yeadon, W. B. *Yeadon's Register of LNER Locomotives Volume 18: Gresley K1 & K2, Thomson K1/1 & Peppercorn K1*. 2000.

Young, John and David Tyreman. *The Hughes and Stanier 2-6-0s*. 2009.